W9-BJZ-821

JUST SIX NUMBERS

Other books by Martin Rees include

Gravity's Fatal Attraction: Black Holes in the Universe
with Mitchell Begelman

Before the Beginning: Our Universe and Others

MARTIN REES

JUST SIX NUMBERS

The Deep Forces That Shape the Universe

A Member of the
Perseus Books Group

First published in 1999 in Great Britain
By Weidenfeld & Nicolson

Published by Basic Books,
A Member of the Perseus Books Group

Typeset at The Spartan Press Ltd,
Lymington, Hants

A CIP catalog record for this book is available from the Library of Congress.
ISBN 0–465–03672–4

CONTENTS

LIST OF ILLUSTRATIONS

PREFACE

Astronomy is the oldest numerical science, crucial in ancient times for calendars and navigation. It is now experiencing a surge of discovery. The enhanced focus on time as we enter the new millennium is boosting interest in our cosmic environment. Astronomy is still the science of numbers, and this book is the story of six that are crucial for our universe, and our place in it.

On the blurred boundaries of ancient maps, cartographers wrote 'There be dragons'. After the pioneer navigators had encircled the globe and delineated the main continents and oceans, later explorers filled in the details. But there was no longer any hope of finding a new continent, or any expectation that the Earth's size and shape would ever be drastically reappraised.

At the end of the twentieth century we have, remarkably, reached the same stage in mapping our universe: the grand outlines are now coming into focus. This is the collective achievement of thousands of astronomers, physicists and engineers, using many different techniques. Modern telescopes probe deep into space; because the light from distant objects takes a long time journeying towards us, they also give us glimpses of the remote past; we have detected 'fossils' laid down in the first few seconds of cosmic history. Spacecraft have revealed neutron stars, black holes, and other extreme phenomena that extend our knowledge of the physical laws. These advances have vastly stretched our cosmic horizons.

There has, in parallel, been an exploration of the microworld within the atom, yielding new insights into the nature of space on the tiniest of scales.

The picture that emerges – a map in time as well as in space – is not what most of us expected. It offers a new perspective on how a single 'genesis event' created billions of galaxies, black holes, stars and planets, and how atoms have been assembled – here on Earth, and perhaps on other worlds – into living beings intricate enough to ponder their origins. There are deep connections between stars and atoms, between the cosmos and the microworld. This book describes – without technicalities – the forces that control us and, indeed, our entire universe. Our emergence and survival depend on very special 'tuning' of the cosmos – a cosmos that may be even vaster than the universe that we can actually see.

ACKNOWLEDGEMENTS

My first debt is to the colleagues with whom I have researched and studied over the years. But I'm equally grateful for the penetrating cosmological discussions I've had with non-specialists: these always bring a fresh perspective, by highlighting the 'big picture', and reminding me that the most important questions are still unanswered. Special thanks, therefore, to David Hart, Graeme Mitchison, Hans Rausing, and Nick Webb. The present book is intended for general readers of this kind. I've tried, while avoiding technicalities, to set new discoveries in context, to distinguish well-based claims from speculations, and to emphasise the mysteries that lie beyond.

I thank John Brockman for inviting me to contribute to the Science Masters series, and for his forbearance during the book's slow gestation. Toby Mundy and Emma Baxter of Weidenfeld and Nicolson were very supportive throughout the editing and production process. I'm grateful to them; also to Richard Sword and Joop Schaye for preparing the illustrations, to Brian Amos for the index, and to Judith Moss for secretarial help.

CHAPTER 1

THE COSMOS AND THE MICROWORLD

> Man is . . . related inextricably to all reality, known and unknowable . . . plankton, a shimmering phosphorescence on the sea and the spinning planets and an expanding universe, all bound together by the elastic string of time. It is advisable to look from the tide pool to the stars and then back to the tide pool again.
>
> John Steinbeck, *The Log from the Sea of Cortez*

SIX NUMBERS

Mathematical laws underpin the fabric of our universe – not just atoms, but galaxies, stars and people. The properties of atoms – their sizes and masses, how many different kinds there are, and the forces linking them together – determine the chemistry of our everyday world. The very existence of atoms depends on forces and particles deep inside them. The objects that astronomers study – planets, stars and galaxies – are controlled by the force of gravity. And everything takes place in the arena of an expanding universe, whose properties were imprinted into it at the time of the initial Big Bang.

Science advances by discerning patterns and regularities in nature, so that more and more phenomena can be subsumed into general categories and laws. Theorists aim to encapsulate the essence of the physical laws in a unified set of equations, and a few numbers. There is still some way to go, but progress is remarkable.

This book describes six numbers that now seem especially significant. Two of them relate to the basic forces; two fix the size and overall 'texture' of our universe and determine whether it will continue for ever; and two more fix the properties of space itself:

- The cosmos is so vast because there is one crucially important huge number $\mathcal{N}$ in nature, equal to 1,000,000, 000,000,000,000,000,000,000,000,000,000. This number measures the strength of the electrical forces that hold atoms together, divided by the force of gravity between them. If $\mathcal{N}$ had a few less zeros, only a short-lived miniature universe could exist: no creatures could grow larger than insects, and there would be no time for biological evolution.

- Another number, ε, whose value is 0.007, defines how firmly atomic nuclei bind together and how all the atoms on Earth were made. Its value controls the power from the Sun and, more sensitively, how stars transmute hydrogen into all the atoms of the periodic table. Carbon and oxygen are common, whereas gold and uranium are rare, because of what happens in the stars. If ε were 0.006 or 0.008, we could not exist.

- The cosmic number Ω (omega) measures the amount of material in our universe – galaxies, diffuse gas, and 'dark matter'. Ω tells us the relative importance of gravity and expansion energy in the universe. If this ratio were too high relative to a particular 'critical' value, the universe would have collapsed long ago; had it been too low, no galaxies or stars would have formed. The initial expansion speed seems to have been finely tuned.

- Measuring the fourth number, λ (lambda), was the biggest scientific news of 1998. An unsuspected new force – a cosmic 'antigravity' – controls the expansion of our universe, even though it has no discernible effect on scales less than a billion light-years. It is destined to become ever more dominant over gravity and other forces as our universe becomes ever darker and emptier. Fortunately for us (and very surprisingly to theorists), λ is very small. Otherwise its effect would have stopped galaxies and stars from forming, and cosmic evolution would have been stifled before it could even begin.

- The seeds for all cosmic structures – stars, galaxies and clusters of galaxies – were all imprinted in the Big Bang. The fabric of our universe depends on one number, $\mathcal{Q}$, which represents the ratio of two fundamental energies and is about 1/100,000 in value. If $\mathcal{Q}$ were even smaller, the universe would be inert and structureless; if $\mathcal{Q}$ were much larger, it would be a violent place, in which no stars or solar systems could survive, dominated by vast black holes.
- The sixth crucial number has been known for centuries, although it's now viewed in a new perspective. It is the number of spatial dimensions in our world, $\mathcal{D}$, and equals three. Life couldn't exist if $\mathcal{D}$ were two or four. Time is a fourth dimension, but distinctively different from the others in that it has a built-in arrow: we 'move' only towards the future. Near black holes, space is so warped that light moves in circles, and time can stand still. Furthermore, close to the time of the Big Bang, and also on microscopic scales, space may reveal its deepest underlying structure of all: the vibrations and harmonies of objects called 'superstrings', in a ten-dimensional arena.

Perhaps there are some connections between these numbers. At the moment, however, we cannot predict any one of them from the values of the others. Nor do we know whether some 'theory of everything' will eventually yield a formula that interrelates them, or that specifies them uniquely. I have highlighted these six because each plays a crucial and distinctive role in our universe, and together they determine how the universe evolves and what its internal potentialities are; moreover, three of them (those that pertain to the large-scale universe) are only now being measured with any precision.

These six numbers constitute a 'recipe' for a universe. Moreover, the outcome is sensitive to their values: if any one of them were to be 'untuned', there would be no stars and no life. Is this tuning just a brute fact, a coincidence? Or is it the providence of a benign Creator? I take the view that it is neither. An infinity of other universes may well exist where the numbers are different. Most would be stillborn or sterile. We could only have emerged (and therefore we naturally now find ourselves) in a universe with the 'right' combination. This realization offers a radically new perspective on our universe, on our place in it, and on the nature of physical laws.

It is astonishing that an expanding universe, whose starting point is so 'simple' that it can be specified by just a few numbers, can evolve (if these numbers are suitable 'tuned') into our intricately structured cosmos. Let us first set the scene by viewing these structures on all scales, from atoms to galaxies.

THE COSMOS THROUGH A ZOOM LENS

Start with a commonplace 'snapshot' – a man and woman – taken from a distance of a few metres. Then imagine the same scene from successively more remote viewpoints, each ten times further away than the previous one. The second frame shows the patch of grass on which they are reclining; the third shows that they are in a public park; the fourth reveals some tall buildings; the next shows the whole city; and the next-but-one a segment of the Earth's horizon, viewed from so high up that it is noticeably curved. Two frames further on, we encounter a powerful image that has been familiar since the 1960s: the entire Earth – continents, oceans, and clouds – with its biosphere seeming no more than a delicate glaze and contrasting with the arid features of its Moon.

Three more leaps show the inner Solar System, with the Earth orbiting the Sun further out than Mercury and Venus; the next shows the entire Solar System. Four frames on (a view from a few light-years away), our Sun looks like a star among its neighbours. After three more frames, we see the billions of similar stars in the flat disc of our Milky Way, stretching for tens of thousands of light-years. Three more leaps reveal the Milky Way as a spiral galaxy, along with Andromeda. From still further, these galaxies seem just two among hundreds of others – outlying members of the Virgo Cluster of galaxies. A further leap shows that the Virgo Cluster is itself just one rather modest cluster. Even if our imaginary telephoto lens had the power of the Hubble Space Telescope, our entire galaxy would, in the final frame, be a barely detectable smudge of light several billion light-years distant.

The series ends there. Our horizon extends no further, but it has taken twenty-five leaps, each by a factor of ten, to reach the limits of our observable universe starting with the 'human' scale of a few metres.

The other set of frames zooms inward rather than outward. From less than one metre, we see an arm; from a few centimetres – as close as we can look with the unaided eye – a small patch of skin. The next frames take us into the fine textures of human tissue, and then into an individual cell (there are a hundred times more cells in our body than there are stars in our galaxy). And then, at the limits of a powerful microscope, we probe the realm of individual molecules: long, tangled strings of proteins, and the double helix of DNA.

The next 'zoom' reveals individual atoms. Here the fuzziness of quantum effects comes in: there is a limit to the sharpness of the pictures we can get. No real microscope can probe within the atom, where a swarm of electrons surrounds the positively charged nucleus, but substructures one hundred times smaller than atomic nuclei can be probed by studying what happens when other particles, accelerated to speeds approaching that of light, are crashed into them. This is the finest detail that we can directly measure; we suspect, however, that the underlying structures in nature may be 'superstrings' or 'quantum foam' on scales so tiny that they would require seventeen more zooms to reveal them.[1]

Our telescopes reach out to a distance that is bigger than a superstring (the smallest substructure postulated to exist within atoms) by a sixty-figure number: there would be sixty frames (of which present measurements cover forty-three) in our 'zoom lens' depiction of the natural world. Of these, our ordinary experience spans nine at most – from the smallest things our eyes can see, about a millimetre in size, to the distance logged on an intercontinental flight. This highlights something important and remarkable, which is so obvious that we take it for granted: our universe covers a vast range of scales, and an immense variety of structures, stretching far larger, and far smaller, than the dimensions of everyday sensations.

LARGE NUMBERS AND DIVERSE SCALES

We are each made up of between 10^{28} and 10^{29} atoms. This 'human scale' is, in a numerical sense, poised midway between the masses of atoms and stars. It would take roughly as many human bodies to make up the mass of the Sun as there are atoms in each of us. But our Sun is just an ordinary star in the galaxy that contains a hundred billion stars altogether. There are at least as many galaxies in our observable universe as there are stars in a galaxy. More than 10^{78} atoms lie within range of our telescope.

Living organisms are configured into layer upon layer of complex structure. Atoms are assembled into complex molecules; these react, via complex pathways in every cell, and indirectly lead to the entire interconnected structure that makes up a tree, an insect or a human. We straddle the cosmos and the microworld – intermediate in size between the Sun, at a billion metres in diameter, and a molecule at a billionth of a metre. It is actually no coincidence that nature attains its maximum complexity on this intermediate scale: anything larger, if it were on a habitable planet, would be vulnerable to breakage or crushing by gravity.

We are used to the idea that we are moulded by the microworld: we are vulnerable to viruses a millionth of a metre in length, and the minute DNA double-helix molecule encodes our total genetic heritage. And it's just as obvious that we depend on the Sun and its power. But what about the still vaster scales? Even the nearest stars are millions of times further away than the Sun, and the known cosmos extends a billion times further still. Can we understand why there is so much beyond our Solar System? In this book I shall describe several ways in which we are linked to the stars, arguing that we cannot understand our origins without the cosmic context.

The intimate connections between the 'inner space' of the subatomic world and the 'outer space' of the cosmos are illustrated by the picture in Figure 1.1 – an *ouraborus*, described by *Encyclopaedia Britannica* as the 'emblematic serpent of ancient Egypt and Greece, represented with its tail in its mouth continually devouring itself and being reborn from itself . . . [It] expresses the unity of all things, material and spiritual, which never disappear but perpetually change form in an eternal cycle of destruction and re-creation'.

On the left in the illustration are the atoms and subatomic particles; this is the 'quantum world'. On the right are planets, stars and galaxies. This book will highlight some remarkable interconnections between the microscales on the left and the macroworld on the right. Our everyday world is determined by atoms and how they combine together into molecules, minerals and living cells. The way stars shine depends on the nuclei within those atoms. Galaxies may be held together by the gravity of a huge swarm of subnuclear particles. Symbolized 'gastronomically' at the top, is the ultimate synthesis that still eludes us – between the cosmos and the quantum.

Lengths spanning sixty powers of ten are depicted in the *ouraborus*. Such an enormous range is actually a prerequisite

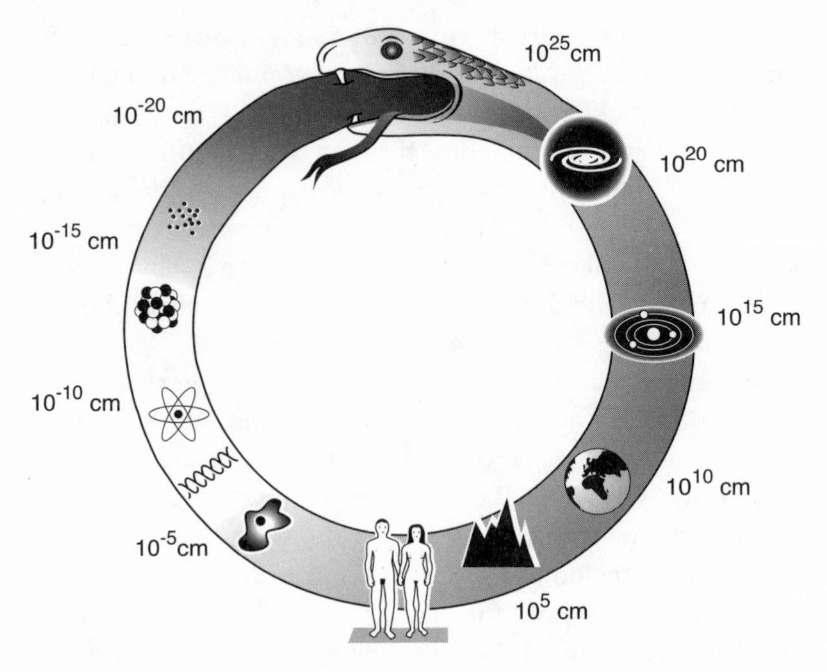

FIGURE 1.1
The *ouraborus*. There are links between the microworld of particles, nuclei and atoms (left) and the cosmos (right).

for an 'interesting' universe. A universe that didn't involve large numbers could never evolve a complex hierarchy of structures: it would be dull, and certainly not habitable. And there must be long timespans as well. Processes in an atom may take a millionth of a billionth of a second to be completed; within the central nucleus of each atom, events are even faster. The complex processes that transform an embryo into blood, bone and flesh involve a succession of cell divisions, coupled with differentiation, each involving thousands of intricately orchestrated regroupings and replications of molecules; this activity never ceases as long as we eat and breathe. And our life is just one generation in humankind's evolution, an episode that is itself just one stage in the emergence of the totality of life.

The tremendous timespans involved in evolution offer a new perspective on the question 'Why is our universe so big?' The emergence of human life here on Earth has taken 4.5

billion years. Even before our Sun and its planets could form, earlier stars must have transmuted pristine hydrogen into carbon, oxygen and the other atoms of the periodic table. This has taken about ten billion years. The size of the observable universe is, roughly, the distance travelled by light since the Big Bang, and so the present visible universe must be around ten billion light-years across.

This is a startling conclusion. The very hugeness of our universe, which seems at first to signify how unimportant we are in the cosmic scheme, is actually entailed by our existence! This is not to say that there couldn't have been a smaller universe, only that we could not have existed in it. The expanse of cosmic space is not an extravagant superfluity; it's a consequence of the prolonged chain of events, extending back before our Solar System formed, that preceded our arrival on the scene.

This may seem a regression to an ancient 'anthropocentric' perspective – something that was shattered by Copernicus's revelation that the Earth moves around the Sun rather than vice versa. But we shouldn't take Copernican modesty (sometimes called the 'principle of mediocrity') too far. Creatures like us require special conditions to have evolved, so our perspective is bound to be in some sense atypical. The vastness of our universe shouldn't surprise us, even though we may still seek a deeper explanation for its distinctive features.

CAN WE HOPE TO UNDERSTAND OUR UNIVERSE?

The physicist Max Planck claimed that theories are never abandoned until their proponents are all dead – that science advances 'funeral by funeral'. But that's too cynical. Several long-running cosmological debates have now been settled; some earlier issues are no longer controversial. Many of us have often changed our minds – I certainly have. Indeed, this book presents a story that I would once myself have thought surprising. The cosmic perspective I'll describe is widely shared, even though many would not go the whole way with my interpretation.

Cosmological ideas are no longer any more fragile and evanescent than our theories about the history of our own Earth. Geologists infer that the continents are drifting over the globe, about as fast as your fingernails grow, and that Europe and North America were joined together 200 million years ago. We believe them, even though such vast spans of time are hard to grasp. We also believe, at least in outline, the story of how our biosphere evolved and how we humans emerged. But some key features of our *cosmic* environment are now underpinned by equally firm data. The empirical support for a Big Bang ten to fifteen billion years ago is as compelling as the evidence that geologists offer on our Earth's history. This is an astonishing turnaround: our ancestors could weave theories almost unencumbered by facts, and until quite recently cosmology seemed little more than speculative mathematics.

A few years ago, I already had ninety per cent confidence that there was indeed a Big Bang – that everything in our observable universe started as a compressed fireball, far hotter than the centre of the Sun. The case now is far stronger: dramatic advances in observations and experiments have brought the broad cosmic picture into sharp focus during the 1990s, and I would now raise my degree of certainty to ninety-nine per cent.

'The most incomprehensible thing about the universe is that it is comprehensible' is one of Einstein's best-known aphorisms, expressing his amazement that the laws of physics, which our minds are somehow attuned to understand, apply not just here on Earth but also in the remotest galaxy. Newton taught us that the same force that makes apples fall holds the Moon and planets in their courses. We now know that this same force binds the galaxies, pulls some stars into black holes, and may eventually cause the Andromeda galaxy to collapse on top of us. Atoms in the most distant galaxies are identical to those we can study in our laboratories. All parts of the universe seem to be evolving in a similar way, as though they shared a common origin. Without this uniformity, cosmology would have got nowhere.

Recent advances bring into focus new mysteries about the origin of our universe, the laws governing it, and even its eventual fate. These pertain to the first tiny fraction of a second after the Big Bang, when conditions were so extreme that the relevant physics isn't understood – where we wonder about the nature of time, the number of dimensions, and the origin of matter. In this initial instant, everything was squeezed to such immense densities that (as symbolized in the *ouraborus*) the problems of the cosmos and the microworld overlap.

Space can't be indefinitely divided. The details are still mysterious, but most physicists suspect that there is some kind of granularity on a scale of 10^{-33} centimetres. This is twenty powers of ten smaller than an atomic nucleus: as big a decrease – as many frames in our 'zoom lens' depiction – as the increase in scale from an atomic nucleus to a major city. We then encounter a barrier: even if there were still tinier structures, they would transcend our concepts of space and time.

What about the largest scales? Are there domains whose light has not yet had time to reach us in the ten billion years or so since the Big Bang? We plainly have no direct evidence. However, there are no theoretical bounds on the extent of our universe (in space, and in future time), and on what may come into view in the remote future – indeed, it may stretch not just millions of times further than our currently observable domain, but *millions of powers of ten* further. And even that isn't all. Our universe, extending immensely far beyond our present horizon, may itself be just one member of a possibly infinite ensemble. This 'multiverse' concept, though speculative, is a natural extension of current cosmological theories, which gain credence because they account for things that we *do* observe. The physical laws and geometry could be different in other universes, and this offers a new perspective on the seemingly special values that the six numbers take in ours.

CHAPTER 2

OUR COSMIC HABITAT I: PLANETS, STARS AND LIFE

> Damn the Solar System. Bad light; planets too distant; pestered with comets; feeble contrivance; could make a better one myself.
>
> Lord Jeffrey

PROTOPLANETS

There is a great cloud in the constellation of Orion, containing enough atoms to make ten thousand Suns. Part of it is a glowing nebula, heated by bright blue stars; the rest is cold, dark and dusty. Within it are warm blobs, emitting no light but generating heat that can be picked up by telescopes fitted with infrared detectors. These blobs are each destined to become stars but are at present 'protostars', contracting under their own gravity. Each is encircled by a disc of gas and dust.

These discs are not unexpected. The dusty cloud in Orion, though denser than most of the expanses between the stars, is still very rarefied, and to form a star some of this gas must contract so much that its density rises a billion billion times. Any slight spin would be amplified during a collapse (a cosmic version of the 'spin-up' when ice skaters pull in their arms) until centrifugal forces prevent all the material from joining the star. Surplus material would be left behind, spinning around each newly formed star. The resultant discs are the precursors of planetary systems: dust particles would collide frequently, sticking together to build up rocky lumps; these in turn would coalesce into larger bodies, which merge to make planets. Our Solar System formed in this way, from a 'protosolar disc'; other stars formed similarly to our Sun, and there is every reason to expect them also to be orbited by retinues of planets.

This scenario, supported by the actual evidence of discs around newly formed stars, has superseded the 'catastrophist' theories popular at the beginning of the twentieth century, which envisaged planetary formation as a rare and special accident. It was thought that our Sun underwent a close encounter with another star – a freakishly rare event because the stars are, on average, so widely dispersed from each other – and that the star's gravitational pull extracted a plume of gas from the Sun; the plume supposedly condensed into 'beads' that each became a planet.

Astronomers in earlier centuries, however, were no more averse than we are today to the idea of other Solar Systems. Back in 1698 Christiaan Huygens, a Dutch scientist who did pioneering work in optics, wrote 'Why [should] not every one of these stars and suns have as great a retinue as our sun, of planets, with their moons to wait upon them?'

OTHER SOLAR SYSTEMS?

Fully formed planets orbiting other stars are harder to detect than their precursor discs. A real highlight of the late 1990s has been the first compelling evidence that planets are indeed common. The principle here is very simple. An observer viewing our Sun from a distance of (say) forty light-years couldn't see any of the planets orbiting it, even with the use of a telescope as powerful as the largest we have on Earth. Nevertheless, the existence of Jupiter (the heaviest planet) could be inferred by careful measurements of the Sun's light.

This is because the Sun and Jupiter are both pivoting around their centre of mass, the so-called 'barycentre'. The Sun is 1,047 times more massive than Jupiter. The barycentre is closer by just that factor to the Sun's centre than to Jupiter's (it actually lies beneath the Sun's surface); the Sun consequently moves about a thousand times more slowly than Jupiter does. Its actual motion is more complicated, because of extra wobbles induced by the other planets, but Jupiter is much the heaviest planet and exerts the dominant effect. By analysing the light very carefully, astronomers have detected small 'wobbles' in the motion of other stars, which are induced by orbiting planets, just as Jupiter induces such motions in our Sun.

The spectrum of starlight reveal patterns due to the distinctive colours emitted or absorbed by the various kinds of atom (carbon, sodium, etc) that stars are made of. If a star moves away from us, its light shifts towards the red end of the spectrum, as compared from the colours emitted by the same atoms in the laboratory – this is the well-known Doppler effect (the analogue, for light, of the way the sound from a receding siren shifts to a lower pitch). If the star is approaching, there is a shift to the blue end of the spectrum. In 1995, two astronomers at the Geneva Observatory, Michel Mayor and Didier Queloz, discovered that the Doppler shift in 51 Pegasi, a nearby star resembling our Sun, was going up and down very slightly as though it was moving in a circle: coming towards us, then receding, then approaching again, and so on in a regular fashion. The implied speed was about fifty metres per second. They inferred that a planet about the size of Jupiter was orbiting it, causing the star to pivot around the centre of mass of the combined system. If the invisible planet were one-thousandth of the star's mass, its orbital speed would be fifty kilometres per second – a thousand times faster than the star is moving.

Geoffrey Marcy and Paul Butler, working in California, have been the champion planet-hunters of the late 1990s. Their instruments can record wavelength shifts of less than one part in a hundred million; they can thereby measure the Doppler effect even when the speeds are only one hundred-millionth of the speed of light – three metres per second – and they have found evidence for planets around many stars. These inferred planets are all big ones, like Jupiter. But this merely reflects the limited sensitivity of their measurements. An Earth-like planet, weighing a few hundred times less than Jupiter, would induce motions of only a few *centimetres* per second, and the Doppler shift would then be only about one part in ten billion – too small to be discerned by the techniques that have discovered the bigger planets.[1]

It should be noted in passing that the telescopes used by the planet-seekers are of moderate size, with mirrors only about two metres in diameter. It is gratifying – and sometimes obscured by the hype that accompanies the biggest projects – that not all important discoveries demand the largest and most expensive equipment. Persistent and ingenious scientists can still achieve a lot with innovative but modest instruments on the ground.

The actual layout of our Solar System is the outcome of many 'accidents'. Rocky asteroids whose orbits cross the Earth's still pose a genuine threat. For example, the impact of a ten-kilometre asteroid, leaving a huge undersea crater near Chicxulub in the Gulf of Mexico, had worldwide climatic effects that probably sealed the dinosaurs' fate sixty-five million years ago; and smaller impacts, still severe enough to cause local devastation, have been more common. But impacts were far more frequent when the Solar System was young, because most of the original protoplanetary bodies within it have by now either been destroyed or kicked out. Our Moon was torn from Earth by a collision with another protoplanet – the intense cratering on its surface bears witness to the violence of its early history. Uranus probably underwent a shattering oblique collision soon after it formed; it is otherwise hard to understand why it spins around an axis almost in the plane of its orbit, in contrast to the other planets, whose axes are more or less aligned perpendicular to that plane.

Pictures beamed back by artificial space probes reveal that all the planets of our own Solar System (and some of their larger moons) are highly distinctive worlds.

It's unlikely that other planetary systems would have the same number of planets, in the same configurations, as our Solar System. Several of those already found have a large Jupiter-like planet closer to the parent star than Mercury (the innermost member of our Solar System) is to our Sun. This is partly an observational bias: heavy planets in fast short-period orbits are easier to detect. The heavy planets already detected may well be accompanied by smaller Earth-like ones.

Only rather special planets could harbour life that in any way resembled what we have on Earth. Gravity must pull strongly enough to prevent their atmosphere from evaporating into space (as would have happened to an atmosphere on our Moon, if it ever had one). For water to exist on their surfaces, planets must be neither too hot nor too cold, and therefore the right distance from a long-lived and stable star. Their orbits must be stable (which they would not be if, for instance, their path was repeatedly crossed by a Jupiter-like planet in an eccentric orbit). The high 'hit rate' of the planet-seekers suggests that there are planets around a high proportion of Sun-like stars in our galaxy. Among these billions of candidates, it would be astonishing if there were not many planets resembling the young Earth.

In the US, NASA's somewhat messianic chief executive, Dan Goldin, has urged that the quest for Earth-like planets – a quest to actually make an image of them rather than just infer them indirectly – should become a main thrust of the space programme. Mere detection of such a faint speck – in Carl Sagan's phrase, a 'pale blue dot' – is a challenge that may take fifteen years to meet. Large arrays of telescopes would have to be deployed in space.

The dim light from a distant world conveys information about cloud cover, the nature of its surface (land or oceans), and perhaps daily or seasonal changes. From the spectrum of the planet's light, we could infer what gases existed in its atmosphere. Our Earth's atmosphere is rich in oxygen; it didn't start out that way, but was transformed by primitive bacteria in its early history. The most interesting question, of course, is whether this may have happened elsewhere: even when a planet offers a propitious environment, what is the chance that simple *organisms* emerge and create a biosphere?

FROM MATTER TO LIFE

Only in the last five years of this millennium have we learnt for sure that there are worlds in orbit around other stars. But we are still little closer to knowing whether any of them harbours anything alive. This question is one for biologists, not for astronomers. It is much more difficult to answer, and there seems no consensus among the experts.

Life on earth has occupied an immense variety of niches. The ecosystems near hot sulphurous outwellings in the deep ocean bed tell us that not even sunlight is essential. We still don't know how or where life got started. A torrid volcano is now more favoured than Darwin's 'warm little pond'; but it could have happened deep underground, or even in dusty molecular clouds in space.

Nor do we know what the odds were against it happening here on Earth – whether life's emergence is 'natural', or whether it involves a chain of accidents so improbable that nothing remotely like it has happened on another planet anywhere else in our galaxy. That's why it would be so crucial to detect life, even in simple and vestigial forms, elsewhere in our Solar System. Mars is still, as it has been since the nineteenth century, the main focus of attention: during the coming years, an armada of space probes is being launched toward the 'red planet' to analyse its surface, to fly over it, and (in later missions) to return samples to Earth. Life could also exist in the ice-covered oceans of Jupiter's frozen moons,

Europa and Callisto, and there are plans to land a submersible probe that could explore beneath the ice.

If life had emerged twice within our Solar System, this would suggest that the entire galaxy would be teeming with life, at least in simple forms. Such a momentous conclusion would require that the two origins were independent. That is an important proviso – for instance, if meteorites from Mars could impact the earth, maybe we are all Martians; conversely, Mars could have been seeded by reverse traffic from Earth!

FROM SIMPLE LIFE TO INTELLIGENCE

We know, at least in outline, the elaborate history and the contingencies that led to our emergence here. For a billion years, primitive organisms exhaled oxygen, transforming the young Earth's poisonous atmosphere and clearing the way for multicellular life. The fossil record tells us that a cornucopia of swimming and creeping things evolved during the Cambrian era 550 million years ago. The next 200 million years saw the greening of the land, offering a habitat for exotic fauna – dragonflies as big as seagulls, millipedes a metre long, scorpions and amphibians. And then the dinosaurs, whose traditional dim and torpid image has been replaced by the dynamism portrayed (in accordance with current scientific opinion) in films such as *Jurassic Park*. They were wiped out in the most sudden and unpredictable of all extinctions: an asteroid crashed onto Earth, causing huge tidal waves and throwing up dust that darkened the sky for years. This opened the way for the line of mammalian descent that led to humans.

Even if we knew that primitive life was widespread, the issue of *intelligent* life would still remain open. An extraordinary procession of species (almost all now extinct) have swum, crawled and flown through our biosphere during its long history. We are the outcome of time and chance: if evolution were rerun, the outcome would be different. Nothing seems to pre-ordain the emergence of intelligence; indeed, some leading evolutionists believe that, even if simple life were widespread in the cosmos, intelligence could be exceedingly rare. We still understand far too little to assess the odds, but there is no reason for obdurate scepticism.

The amazing and fascinating complexity of biological evolution, and the variety of life on Earth, makes us realize that everything in the inanimate world is, in comparison, very simple. And this simplicity – or, at least, *relative* simplicity – is a feature of the objects that astronomers study. Things are hard to understand because they are *complex*, not because they are *big*. The challenge of fully elucidating how atoms assembled themselves – here on Earth, and perhaps on other worlds – into living beings intricate enough to ponder their origins is more daunting than anything in cosmology. For just that reason, I don't think it's presumptuous to aspire to understand our large-scale universe.

The concept of a 'plurality of inhabited worlds' is still the province of speculative thinkers, as it has been through the ages. The year 2000 marks the fourth centenary of the death of Giordano Bruno, burnt at the stake in Rome. He believed that:

> In space there are countless constellations, suns and planets; we see only the suns because they give light; the planets remain invisible, for they are small and dark. There are also numberless earths circling around their suns, no worse and no less than this globe of ours. For no reasonable mind can assume that heavenly bodies that may be far more magnificent than ours would not bear upon them creatures similar or even superior to those upon our human earth.

Ever since Bruno's time, this belief has been widely shared. In the eighteenth century, the great astronomer William Herschel, discoverer of the planet Uranus, thought that the planets, the Moon, and even the Sun were inhabited. In the 1880s, Percival Lowell, a wealthy American, built his own observatory in Flagstaff, Arizona, primarily to study Mars. He believed that the 'canals' (now recognized to be no more than a combination of wishful thinking and optical illusion) were an irrigation project to channel water from the frozen polar caps to the 'deserts' of its equatorial zones. In 1900, a French foundation offered the Guzman Prize of 100,000 francs for the first contact with an extraterrestrial species; but prudence led them to exclude Mars – detecting Martians was thought to be too easy!

A COMMON CULTURE WITH ALIENS?

Searches for extraterrestrial intelligence (SETI) are being spearheaded by scientists at the SETI Institute in Mountain View, California. The efforts have concentrated on searches for radio transmissions that could be artificial in origin, and have used various large radio telescopes around the world. This option is familiar also from fictional depictions such as Carl Sagan's *Contact* (in which it generally pays off). But radio is not the only conceivable channel: narrow-beamed lasers could span interstellar distances with a modest power consumption. We already have the technology, if we so wish, to proclaim our presence many light-years away by either of these methods; indeed, the combined effects of all radio transmitters, radars and so forth would in any case reveal us to any aliens with sensitive radio telescopes. We know so little about the origin and potentialities of life that it is hard to assess what method for detecting it is best. So it is sensible to use every available technique and be alert to all possibilities. But we should be mindful of 'observational selection': even if we do discover something, we can't infer that it is 'typical', because our instruments and techniques restrict us to detecting a biased and incomplete selection of what may actually be out there.

There may be no other intelligent life elsewhere. Even if there is, it may be on some water-covered world where super-dolphins enjoy a contemplative oceanic life, doing nothing to reveal themselves. There are heavy odds against success, but systematic scans for artificial signals are a worthwhile gamble because of the philosophical import of *any* detection. A manifestly artificial signal – even if it were as boring as lists of prime numbers, or the digits of 'pi' – would imply that 'intelligence' wasn't unique to the Earth and had evolved elsewhere. The nearest potential sites are so far away that signals would take many years in transit. For this reason alone, transmission would be primarily one-way. There would be time to send a measured response, but no scope for quick repartee!

Any remote beings who could communicate with us would have some concepts of mathematics and logic that paralleled our own. And they would also share a knowledge of the basic particles and forces that govern our universe. Their habitat may be very different (and the biosphere even more different) from ours here on Earth; but they, and their planet, would be made of atoms just like those on Earth. For them, as for us, the most important particles would be protons and electrons: one electron orbiting a proton makes a hydrogen atom, and electric currents and radio transmitters involve streams of electrons. A proton is 1,836 times heavier than an electron, and the number 1,836 would have the same connotations to any 'intelligence' able and motivated to transmit radio signals. All the basic forces and natural laws would be the same. Indeed, this uniformity – without which our universe would be a far more baffling place – seems to extend to the remotest galaxies that astronomers can study. (Later chapters in this book will, however, speculate about other 'universes', forever beyond range of our telescopes, where different laws may prevail.)

Clearly, alien beings wouldn't use metres, kilograms or seconds. But we could exchange information about the *ratios* of two masses (such as the ratio of proton and electron masses) or of two lengths, which are 'pure numbers' that don't depend on what units are used: the statement that one rod is ten times as long as another is true (or false) whether we measure lengths in feet or metres or some alien units. As Richard Feynman noted, he could tell extraterrestrials that he was 'seventeen billion hydrogen atoms high' and they should understand him.

Some 'intelligences' could exist with no intellectual affinity to us whatsoever. But any beings who transmitted a signal to us must have achieved some mastery over their physical surroundings. If they had any powers of reflection, they would surely share our curiosity about the cosmic 'genesis event' from which we've all emerged. They would be likely to be interested in how our universe is structured into stars and galaxies, what it contains, how it is expanding, and its eventual destiny. These things would be part of the common culture that we would share with any aliens. They would note, as we do, that a few key numbers are crucial to our shared cosmic environment.

Six of these numbers are the theme of the present book. They determine key features of our universe: how it expands; whether planets, stars and galaxies can form; and whether there can be a 'chemistry' propitious for evolution. Moreover, the nature of our universe is remarkably *sensitive* to these numbers. If you imagine setting up a universe by adjusting six dials, then the tuning must be precise in order to yield a universe that could harbour life. Is this providence? Is it coincidence? Are these numbers the outcome of a 'theory of everything' that uniquely fixes them? None of these interpretations seems compelling. Instead, I believe that the apparent 'tuning' intimates something even more remarkable: that our observable universe – all we can see out to the limits of our telescopes – is just one part of an ensemble, among which there is even a diversity of physical laws. This is speculation, but it is compatible with the best theories we have.

We know that there are planets orbiting other stars, just as the Earth orbits our own star, the Sun. We may wonder what habitats they offer. Is their gravity too weak to retain an atmosphere? Are they too hot, too cold, or too dry to harbour life? Probably only a few offer an environment conducive for life. So, on a much grander scale, there may be innumerable other universes that we cannot observe because light from them can never reach us. Would they be propitious for the kind of evolution that has happened on at least one planet around at least one star in our 'home' universe? In most of them, the six numbers could be different: only a few universes would then be 'well tuned' for life. We should not be surprised that, in our universe, the numbers seem providentially tuned, any more than we should be surprised to find ourselves on a rather special planet whose gravity can retain an atmosphere, where the temperature allows water to exist, and that is orbiting a stable long-lived star.

CHAPTER 3

THE LARGE NUMBER $\mathcal{N}$: GRAVITY IN THE COSMOS

> Who could believe an ant in theory?
> A giraffe in blueprint?
> Ten thousand doctors of what's possible
> Could reason half the jungle out of being.
> John Ciardi

NEWTON'S 'CLOCKWORK'

If we were establishing a discourse with intelligent beings on another planet, it would be natural to start with gravity. This force grips planets in their orbits and holds the stars together. On a still larger scale, entire galaxies – swarms of billions of stars – are governed by gravity. No substance, no kind of particle, not even light itself escapes its grasp. It controls the expansion of the entire universe, and perhaps its eventual fate.

Gravity still presents deep mysteries. It is more perplexing than any of the other basic forces of nature. But it was the first force to be described in a mathematical fashion. Sir Isaac Newton told us in the seventeenth century that the attraction between any two objects obeys an 'inverse square law'. The force weakens in proportion to the square of the distance between the two masses: take them twice as far away and the attraction between them is four times weaker. Newton

$$F = \frac{K \cdot M_1 \cdot M_2}{d^2}$$

realized that the force that makes apples fall and governs a cannon-ball's trajectory is the same that locks the Moon in its orbit around the Earth. He proved that his law accounted for the elliptical orbits of the planets – a compelling demonstration of the power of mathematics to predict the 'clockwork' of the natural world.

Newton's great work, the *Principia*, published in 1687, is a three-volume Latin text, laced with elaborate theorems of a mainly geometric kind. It is a monument to the pre-eminent scientific intellect of the millennium. Despite the forbidding austerity of his writings (and his personality), Newton's impact was immense, on philosophers and poets alike. And that influence percolated to a wider public as well: for instance, a book entitled *Newtonianism for Ladies* was published in 1737. The essence of his theory of gravity appeared in a more accessible book called *The System of the World*.

In this latter work, a key idea is neatly illustrated by a picture showing cannon-balls fired horizontally from a mountain-top. The faster they're flung, the further they go before hitting the ground. If the speed is very high, the earth 'falls away' under the projectile's trajectory, and it goes into orbit. The requisite speed (about eight kilometres per second) was of course far beyond the cannons of Newton's time, but today we're familiar with artificial satellites that stay in orbit for just this reason. Newton himself showed that the same force holds the planets in their elliptical orbits round the Sun. Gravity acts on a grander scale in clusters of stars; and in galaxies, where billions of stars are held in orbit around a central hub.

In the Sun and other stars like it, there is a balance between gravity, which pulls them together, and the pressure of their hot interior, which, if gravity didn't act, would make them fly apart. In our own Earth's atmosphere, the pressure at ground level, likewise, balances the weight of all the air above us.

GRAVITY ON BIG AND SMALL SCALES

Our Earth's gravity has more drastic effects on big objects than small ones. When producers of 'disaster movies' use a model to depict (for example) a bridge or dam collapsing, they must make it not of real steel and concrete but of very flimsy material that bends or shatters when dropped from table-top height. And the film has to be shot fast and replayed in slow motion to look realistic. Even when this is carefully done, there may be other give-away clues that we are viewing a miniature version rather than the real thing – for instance, small wavelets in a water tank are smoothed by surface tension (the force that holds raindrops together), but this effect is negligible in a full-scale turbulent river or in ocean waves. Surface tension allows spiders to walk on water, but we can't.

Being the right size is crucial in the biological world. Large animals are not just blown-up versions of small ones: they are differently proportioned, with, for instance, thicker legs in relation to their height. Imagine you doubled the dimensions of an animal, but kept its shape the same. Its volume and weight would become eight (2^3) times larger, not just twice as large; but the cross-section of its legs would only go up by a factor of four (2^2) and would be took weak to support it. It would need a redesign. The bigger they are, the harder they fall: 'godzillas' would need legs thicker than their bodies, and would not survive a fall; mice, on the other hand, can climb vertically, and are unharmed even when dropped from many times their own height.

Galileo (who died in the same year that Newton was born) was the first clearly to realize these constraints on size. He wrote:

> Nor could Nature make trees of immeasurable size, because their branches would eventually fall of their own weight . . . When bodies are diminished, their strengths do not proportionally diminish; rather, in very small bodies the strength

> grows in greater ratio, and I believe that a little dog might carry on his back two or three dogs of the same size, whereas I doubt that a horse could carry even one horse his size.

Similar arguments limit the size of birds (the constraints are more stringent for humming birds that can hover than for albatrosses that glide); but the limits are more relaxed for floating creatures, allowing leviathans in the ocean. In contrast, being too small leads to problems of another kind: a large area of skin in proportion to weight, whereby heat is lost quickly; small mammals and birds must eat and metabolize fast in order to stay warm.

There would be analogous limits on other worlds. For example, the physicist Edwin Salpeter has speculated, along with Carl Sagan, on the ecology of hypothetical balloon-like creatures that could survive in the dense atmosphere of Jupiter. Each new generation would face a race against time: it would have to inflate large enough to achieve buoyancy before gravity pulled it to destruction in the dark high-pressure layers deeper down.

THE VALUE OF $\mathcal{N}$ AND WHY IT IS SO LARGE

Despite its importance for us, for our biosphere, and for the cosmos, gravity is actually *amazingly feeble* compared with the other forces that affect atoms. Electric charges of opposite 'sign' attract each other: a hydrogen atom consists of a positively-charged proton, with a single (negative) electron trapped in orbit around it. Two protons would, according to Newton's laws, attract each other gravitationally, as well as exerting an electrical force of repulsion on one another. Both these forces depend on distance in the same way (both follow an 'inverse square' law), and so their relative strength is measured by an important number, $\mathcal{N}$, which is the same irrespective of how widely separated the protons are. When two hydrogen atoms are bound together in a molecule, the electric force between the protons is neutralized by the two electrons. The gravitational attraction between the protons is thirty-six powers of ten feebler than the electrical forces, and quite unmeasurable. Gravity can safely be ignored by chemists when they study how groups of atoms bond together to form molecules.

How, then, can gravity nonetheless be dominant, pinning us to the ground and holding the moon and planets in their courses? It's because gravity is *always an attraction*: if you double a mass, then you double the gravitational pull it exerts. On the other hand, electric charges can repel each other as well as attract; they can be either positive or negative. Two charges only exert twice the force of one if they are of the same 'sign'. But any everyday object is made up of huge numbers of atoms (each made up of a positively charged nucleus surrounded by negative electrons), and the positive and negative charges almost exactly cancel out. Even when we are 'charged up' so that our hair stands on end, the imbalance is less than one charge in a billion billion. But everything has the same sign of 'gravitational charge', and so gravity 'gains' relative to electrical forces in larger objects. The balance of electric forces is only slightly disturbed when a solid is compressed or stretched. An apple falls only when the combined gravity of all the atoms in the Earth can defeat the electrical stresses in the stalk holding it to the tree. Gravity is important to us because we live on the heavy Earth.

We can quantify this. In Chapter 1, we envisaged a set of pictures, each being viewed from ten times as far as the last. Imagine now a set of differently sized spheres, containing respectively 10, 100, 1000, . . . atoms, in other words each ten times heavier than the one before. The eighteenth would be as big as a grain of sand, the twenty-ninth the size of a human, and the fortieth that of a largish asteroid. For each thousand-fold increase in mass, the volume also goes up a thousand times (if the spheres are equally dense) but the radius goes up only by ten times. The importance of the sphere's own gravity, measured by how much energy it takes to remove an atom from its gravitational pull, depends on mass divided by radius,[1] and so goes up a factor of a hundred. Gravity starts off, on the atomic scale, with a handicap of thirty-six powers of ten; but it gains two powers of ten (in other words 100) for every three powers (factors of 1,000) in mass. So gravity will have caught up for the fifty-fourth object ($54 = 36 \times 3/2$), which has about Jupiter's mass. In any still heavier lump more massive than Jupiter, gravity is so strong that it overwhelms the forces that hold solids together.

Sand grains and sugar lumps are, like us, affected by the gravity of the massive Earth. But their *self-gravity* – the gravitational pull that their constituent atoms exert on each other, rather than on the entire Earth – is negligible. Self-gravity is not important in asteroids, nor in Mars's two small potato-shaped moons, Phobos and Deimos. But bodies as large as planets (and even our own large Moon) are not rigid enough to maintain an irregular shape: gravity makes them nearly round. And masses above that of Jupiter get crushed by their own gravity to extraordinary densities (unless the centre gets hot enough to supply a balancing pressure, which is what happens in the Sun and other stars like it). It is because gravity is so weak that a typical star like the Sun is so massive. In any lesser aggregate, gravity could not compete with pressure, nor squeeze the material hot and dense enough to make it shine.

The Sun contains about a thousand times more mass than Jupiter. If it were cold, gravity would squeeze it a million times denser than an ordinary solid: it would be a 'white dwarf' about the size of the Earth but 330,000 times more massive. But the Sun's core actually has a temperature of fifteen million degrees – thousands of times hotter than its glowing surface, and the pressure of this immensely hot gas 'puffs up' the Sun and holds it in equilibrium.

The English astrophysicist Arthur Eddington was among the first to understand the physical nature of stars. He speculated about how much we could learn about them just by theorizing, if we lived on a perpetually cloud-bound planet. We couldn't, of course, guess how many there are, but simple reasoning along the lines I've just outlined could tell us how big they would have to be, and it isn't too difficult to extend the argument further, and work out how brightly such objects could shine. Eddington concluded that: 'When we draw aside the veil of clouds beneath which our physicist is working and let him look up at the sky, there he will find a thousand million globes of gas, nearly all with [these] masses.'

Gravitation is feebler than the forces governing the microworld by the number $\mathcal{N}$, about 10^{36}. What would happen if it weren't quite so weak? Imagine, for instance, a universe where gravity was 'only' 10^{30} rather than 10^{36} feebler than electric forces. Atoms and molecules would behave just as in our actual universe, but objects would not need to be so large before gravity became competitive with the other forces. The number of atoms needed to make a star (a gravitationally bound fusion reactor) would be a billion times less in this imagined universe. Planet masses would also be scaled down by a billion. Irrespective of whether these planets could retain steady orbits, the strength of gravity would stunt the evolutionary potential on them. In an imaginary strong-gravity world, even insects would need thick legs to support them, and no animals could get much larger. Gravity would crush anything as large as ourselves.

Galaxies would form much more quickly in such a universe, and would be miniaturized. Instead of the stars being widely dispersed, they would be so densely packed that close encounters would be frequent. This would in itself preclude stable planetary systems, because the orbits would be disturbed by passing stars – something that (fortunately for our Earth) is unlikely to happen in our own Solar System.

But what would preclude a complex ecosystem even more would be the limited time available for development. Heat would leak more quickly from these 'mini-stars': in this hypothetical strong-gravity world, stellar lifetimes would be a million times shorter. Instead of living for ten billion years, a typical star would live for about 10,000 years. A mini-Sun would burn faster, and would have exhausted its energy before even the first steps in organic evolution had got under way. Conditions for complex evolution would undoubtedly be less favourable if (leaving everything else unchanged) gravity were stronger. There wouldn't be such a huge gulf as there is in our actual universe between the immense time-spans of astronomical processes and the basic microphysical timescales for physical or chemical reactions. The converse, however, is that an even *weaker* gravity could allow even more elaborate and longer-lived structures to develop.

Gravity is the organizing force for the cosmos. We shall see in Chapter 7 how it is crucial in allowing structure to unfold from a Big Bang that was initially almost featureless. But it is only because it is weak compared with other forces that large and long-lived structures can exist. Paradoxically, the weaker gravity is (provided that it isn't actually zero), the grander and more complex can be its consequences. We have no theory that tells us the value of $\mathcal{N}$. All we know is that nothing as complex as humankind could have emerged if $\mathcal{N}$ were much less than 1,000,000,000,000,000,000,000,000,000,000,000,000.

FROM NEWTON TO EINSTEIN

More than two centuries after Newton, Einstein proposed his theory of gravity known as 'general relativity'. According to this theory, planets actually follow the straightest path in a 'space-time' that is curved by the presence of the Sun. It is commonly claimed that Einstein 'overthrew' Newtonian physics, but this is misleading. Newton's law still describes motions in the Solar System with good precision (the most famous discrepancy being a slight anomaly in Mercury's orbit that was resolved by Einstein's theory) and is adequate for programming the trajectories of space probes to the Moon and planets. Einstein's theory, however, copes (unlike Newton's) with objects whose speeds are close to that of light, with the ultra-strong gravity that could induce such enormous speeds, and with the effect of gravity on light itself. More importantly, Einstein *deepened our understanding* of gravity. To Newton, it was a mystery why all particles fell at the same rate and followed identical orbits – why the force of gravity and the inertia were in exactly the same ratio for all substances (in contrast to electric forces, where the 'charge' and 'mass' are not proportionate) – but Einstein showed that this was a natural consequence of all bodies taking the same 'straightest' path in a space-time curved by mass and energy. The theory of general relativity was thus a conceptual breakthrough – especially remarkable because it stemmed from Einstein's deep insight rather than being stimulated by any specific experiment or observation.

Einstein didn't 'prove Newton wrong'; he transcended Newton's theory by incorporating it into something more profound, and with wider applicability. It would actually have been better (and would have obviated widespread misunderstanding of its cultural implications) if his theory had been given a different name: not 'the theory of relativity' but 'the theory of invariance'. Einstein's achievement was to discover a set of equations that can be applied by any observer and incorporate the remarkable circumstance that the speed of light, measured in any 'local' experiment, is the same however the observer is moving.

The development of any science is marked by increasingly general theories, that subsume previously unrelated facts and extend the scope of those that precede them. The physicist and historian Julian Barbour offers a mountaineering metaphor,[2] which I think rings true:

> The higher we climb, the more comprehensive the view. Each new vantage point yields a better understanding of the interconnection of things. What is more, gradual accumulation of understanding is punctuated by sudden and startling enlargements of the horizon, as when we reach the brow of a

> hill and see things never conceived of in the ascent. Once we have found our bearings in the new landscape, our path to the most recently attained summit is laid bare and takes its honourable place in the new world.

Experience shapes our intuition and common sense: we assimilate the physical laws that directly affect us. Newton's laws are in some sense 'hardwired' into monkeys that swing confidently from tree to tree. But far out in space lie environments differing hugely from our own. We should not be surprised that commonsense notions break down over vast cosmic distances, or at high speeds, or when gravity is strong.

An intelligence that could roam rapidly through the universe – constrained by the basic physical laws but not by current technology – would extend its intuitions about space and time to incorporate the distinctive and bizarre-seeming consequences of relativity. The speed of light turns out to have very special significance: it can be approached, but never exceeded. But this 'cosmic speed limit' imposes no bounds to how far you can travel in your lifetime, because clocks run slower (and on-board time is 'dilated') as a spaceship accelerates towards the speed of light. However, were you to travel to a star a hundred light-years away, and then return, more than two hundred years would have passed at home, however young you still felt. Your spacecraft cannot have made the journey faster than light (as measured by a stay-at-home observer), but the closer your speed approached that of light, the less you would have aged.

These effects are counterintuitive simply because our experience is limited to slow speeds. An airliner flies at only a millionth of the speed of light, not nearly fast enough to make the time dilation perceptible: even for the most inveterate air traveller it would be less than a millisecond over an entire lifetime. This tiny effect has, nevertheless, now been measured, and found to accord with Einstein's predictions, by experiments using atomic clocks accurate to a billionth of a second.

A related 'time dilation' is caused by gravity: near a large mass, clocks tend to run slow. This too is almost imperceptible here on Earth because, just as we are only used to 'slow' motions, we experience only 'weak' gravity. This dilation must, however, be allowed for, along with the effects of orbital motion, in programming the amazingly accurate Global Positioning Satellite (GPS) system.

A measure of the strength of a body's gravity is the speed with which a projectile must be fired to escape its grasp. It takes 11.2 kilometres per second to escape from the Earth. This speed is tiny compared with that of light, 300,000 kilometres per second, but it challenges rocket engineers constrained to use chemical fuel, which converts only a billionth of its so-called 'rest-mass energy' (Einstein's mc^2 – see Chapter 4) into effective power. The escape velocity from the Sun's surface is 600 kilometres per second – still only one fifth of one per cent of the speed of light.

'STRONG GRAVITY' AND BLACK HOLES

Newtonian theory works, with only very small corrections, everywhere in our Solar System. But we should prepare for surprises when gravity is far stronger. And astronomers have discovered such places: neutron stars, for instance. Stars leave these ultra-dense remnants behind when they explode as supernovae (discussed further in the next chapter). Neutron stars are typically 1.4 times as massive as the Sun, but only about twenty kilometres across; on their surface, the gravitational force is a million million times fiercer than on Earth. More energy is needed to rise a millimetre above a neutron star's surface than to break completely free of Earth's gravity. A pen dropped from a height of one metre would impact with the energy of a ton of TNT (although the intense gravity on a neutron star's surface would actually, of course, squash any such objects instantly). A projectile would need to attain half the speed of light to escape its gravity; conversely, anything that fell freely onto a neutron star from a great height would impact at more than half the speed of light.

Newton's theory cannot cope when gravity is as powerful as it is around neutron stars; Einstein's general relativity is needed. Clocks near the surface would run ten to twenty per cent slower compared with those far away. Light from the surface would be strongly curved, so that, viewing from afar, you would see not just one hemisphere but part of the backside of the neutron star as well.

A body that was a few times smaller, or a few times heavier, than a neutron star would trap all the light in its vicinity and become a black hole; the space around it would 'close up' on itself. If the Sun were squeezed down to a radius of three kilometres, it would become a black hole. Fortunately, Nature has done such experiments for us, because the cosmos is known to contain objects that have collapsed, 'puncturing' space and cutting themselves off from the external universe.

There are many millions of black holes in our galaxy, of about ten solar masses each, which are the terminal state of massive stars or perhaps the outcome of collisions between stars. When isolated in space, such objects are very inconspicuous: they can be detected only by the gravitational effect that they exert on other bodies or light rays that pass close to them. Easier to detect are those with an ordinary star orbiting around them to make a binary system. The technique is similar to that used to infer planets from the motion they induce in their parent star; but in this case the task is easier because the visible star is of lower mass than the dark object (instead of being a thousand or more times heavier), and so gyrates in a larger and faster orbit.

Astronomers are always specially interested in the most 'extreme' phenomena in the cosmos, because it is through studying these that we are most likely to learn something fundamentally new. Perhaps most remarkable of all are the amazingly intense flashes called 'gamma-ray bursts'. These events, so powerful that for a few seconds they outshine a million entire galaxies of stars, are probably black holes caught in the act of formation.

Much larger black holes lurk in the centres of galaxies. We infer their presence by observing intense radiation from gas swirling around them at close to the speed of light, or by detecting the ultra-rapid motions of stars passing close to them. The stars very close to the centre of our own galaxy are orbiting very fast, as though feeling the gravity of a dark mass: a black hole with a mass of 2.5 million Suns. The size of a black hole is proportionate to its mass, and the hole at the Galactic Centre has a radius of six million kilometres. Some of the real monsters in the centres of other galaxies, weighing as much as several billion suns, are as big as our entire Solar System – although they are nonetheless still very small compared with the galaxies in whose cores they lurk.

Peculiar and counterintuitive though they are, black holes are actually simpler to describe than any other celestial object. The Earth's structure depends on its history, and on what it's made of; similarly sized planets orbiting other stars would assuredly be very different. And the Sun, basically a huge globe of gas exhibiting continuous turbulence and flaring on its surface, would look different if it contained a different 'mix' of atoms. But a black hole loses all 'memory' of how it was formed and quickly settles down to a standard smooth state described just by two quantities: how much mass went into it, and how fast it is spinning. In 1963, long before there was any evidence that black holes existed – before, indeed, the American physicist John Archibald Wheeler introduced the name 'black hole' – a theorist from New Zealand, Roy Kerr, discovered a solution of Einstein's equations that represented a spinning object. Later work by others led to the remarkable result that *anything* that collapses would settle down into a black hole that was exactly described by Kerr's formula. Black holes are as standardized as elementary particles. Einstein's theory tells us exactly how they distort space and time, and what shape their 'surface' is.

Around black holes, our intuitions about space and time go

badly awry. Light travels along the 'straightest' path, but in strongly warped space this can be a complicated curve. And near them, time runs very slowly (even more slowly than near a neutron star). Conversely, if you could hover, or orbit, very close to a black hole, you would see the external universe speeded up. There is a well-defined 'surface' around a black hole, where, to an observer at a safe distance, clocks (or an in-falling experimenter) would seem to 'freeze' because the time dilation becomes almost infinite.

Not even light can escape from inside this surface: the distortions of space and time are even worse. It is as though space itself is being sucked in so fast that even an outwardly directed light ray is dragged inwards. In a black hole, you can no more move 'outwards' in space than you can move backwards in time.

A *spinning* black hole distorts space and time in a more complicated way. To envisage it, imagine a whirlpool in which water spirals towards a central vortex. Far away from the vortex, you can navigate as you wish, either going with the flow or making headway against it. Closer in, the water swirls faster than your boat's speed: you are constrained to go round with the flow, although you can still move outwards (on an outward spiral) as well as in. But, closer still, even the inward flow becomes faster than your boat. If you venture within some 'critical radius' you have no choice about your fate, and are sucked in towards destruction.

A black hole is shrouded by a surface that acts like a one-way membrane. No signals from inside can be transmitted to colleagues watching from a safe distance. Anyone who passes inside the 'surface' is trapped, and fated to be sucked inward towards a region where, according to Einstein's equations, gravity 'goes infinite' within a finite time, as measured by their own clock. This 'singularity' actually signifies that conditions transcend the physics that we know about, just as we believe they did at the very beginning of our universe. Anyone falling into a black hole thus encounters 'the end of time'. Is this a foretaste of the Big Crunch that could be the ultimate fate of our universe? Or does our universe have a perpetual future? Or could some still-unknown physics protect us from this fate?

Einstein's theory was, famously, triggered by his 'happy thought' that gravity was indistinguishable from accelerated motion and would be undetectable in a freely falling lift. *Non-uniformities* in gravity cannot, however, be eliminated. If a phalanx of kamikaze astronauts were in free fall towards the Earth in regular formation, then the horizontal spacings between them would shrink but the vertical spacings would increase. This is because their trajectories all converge towards the centre of the Earth, and the gravitational force pulls more strongly on those lower down in the formation and hence nearer to the Earth. And there would be a similar effect between the different parts of each astronaut's body: falling feet-first, the astronaut would feel a vertical stretching and a sideways compression. This 'tidal' force, imperceptible for astronauts in the Earth's gravity, becomes catastrophically large in a black hole, leading to shredding and 'spaghettification' before the central 'singularity' is reached. An astronaut falling towards a stellar-mass black hole would feel severe tidal effects even before reaching the hole's surface; thereafter, only a few milliseconds would remain (as measured by the astronaut's clock) before encountering the singularity. But tidal effects are more gentle around the supermassive black holes in the centres of galaxies: even after passing inside the surface of one of these, several hours would remain for leisured exploration before getting close enough to the central singularity to be severely discomforted.[3]

ATOMIC-SCALE BLACK HOLES

Black holes are a remarkable theoretical construct, but they are more than that. Evidence that they actually exist is now compelling. They are implicated in some of the most spectacular phenomena we observe in the cosmos – quasars and explosive outbursts. There are still lively debates about exactly how they formed, but there is no mystery about how gravity could have overwhelmed all other forces in a dead star, or within a cloud of gas in the centre of a galaxy. These formation processes require them to be at least as massive as a star, because we've seen that, for asteroids and planets, gravity can't compete with other forces. Indeed, a physicist on a cloud-bound planet could have predicted that if stars existed, then so probably did stellar-mass black holes.

The scale of stars, which determines the mass of black holes that can actually form today, stems, as we've seen, from a balance between gravitational and atomic forces. But nothing in Einstein's theory picks out any special mass. Black holes are made from the fabric of space itself. Insofar as space is a smooth continuum, nothing apart from a simple scaling distinguishes whether a hole (once it has formed) is as big as an atom, or as big as a star, or as big as our observable universe.

Even a hole that was only the size of an atom would have the mass of a mountain. Black holes are, by definition, objects where gravity has overwhelmed all other forces. For an atom-sized black hole to form, 10^{36} atoms must be squeezed into the dimensions of one. This forbidding requirement is another consequence of the hugeness of our cosmic number $\mathcal{N}$, which measures the weakness of gravity on the atomic scale. What about black holes even smaller than an atom? Here there is an eventual limit (which will reappear in Chapter 10) due to an inherent graininess of space on the tiniest scale.

Atomic-scale black holes could have formed, if at all, only in the immense pressures that prevailed in the earliest instants of the universe. If they actually existed, such mini-holes would be extraordinary 'missing links' between the cosmos and the microworld.

CHAPTER 4

STARS, THE PERIODIC TABLE, AND ε

> I believe a leaf of grass is no less than the journey-work of the stars.
>
> Walt Whitman

STARS AS 'NUCLEAR FUSION REACTORS'

How old is the Earth? It is now pinned down, by measurements of radioactive atoms, to 4.55 billion years. Compelling arguments for its great antiquity were, however, already being advanced in the nineteenth century. Geologists, gauging the rate at which erosion and sedimentation shaped our terrain, assessed the Earth's age as at least a billion years: Darwinians concurred with this, from their estimates of how many generations of gradually evolving species must have lived before us. On the other hand, the great physicist Lord Kelvin calculated that all the Sun's internal heat would leak out, and it would deflate, in only one per cent of that time. He gloomily averred that: 'Inhabitants of the Earth cannot continue to enjoy the light and heat essential to their life, for many millions of years longer, unless sources now unknown to us are prepared in the great storehouse of creation.' Twentieth-century science has taught us that such a source indeed exists, stored in the nuclei of atoms. H-bombs are frightening testimony to the energy latent in the nucleus.

The Sun is fuelled by conversion of hydrogen (the simplest atom, whose nucleus consists of one proton) into helium (the second-simplest nucleus, consisting of two protons and two neutrons). Attempts to harness fusion as a power source ('controlled fusion') have so far been stymied by the difficulty of achieving the requisite temperatures of many millions of degrees. It is even more of a problem to confine this ultra-hot gas physically in a laboratory – it would obviously melt any solid container – and it has instead to be trapped by magnetic forces. But the Sun is so massive that gravity holds down the overlying cooler layers, and thereby 'keeps the lid on' the high-pressure core. The Sun has adjusted its structure so that nuclear power is generated in the core, and diffuses outward, at just the rate needed to balance the heat lost from the surface – heat that is the basis for life on Earth.

This fuel has kept the Sun shining for nearly five billion years. But when it starts to run out, in another 5 billion years or so, the Sun's core will contract, and the outer layers expand. For a hundred million years – a brief interval relative to its overall lifetime – the Sun will brighten up and expand into the kind of star known as a 'red giant', engulfing the inner planets and vaporizing any life that remains on Earth. Some of its outer layers will be blown off, but the core will then settle down as a white dwarf, shining with a dull blue glow, no brighter than a full moon today, on the parched remains of the Solar System.

Astrophysicists have computed what the inside of our Sun should be like, and have achieved a gratifying fit with its observed radius, brightness, temperature and so forth. They can tell us confidently what conditions prevail in its deep interior; they can also calculate how it will evolve over the next few billion years. Obviously these calculations can't be checked directly. We can, however, observe other stars *like* the Sun that are at different stages in their evolution. Having a single 'snapshot' of each star's life is not a fatal handicap if we have a large sample, born at different times, available for study. In the same way, a newly landed Martian wouldn't take long to infer the life-cycle of humans (or of trees), by observing large numbers at different stages. Even among the nearby stars, we can discern some that are still youngsters, no more than a million years old, and others in a near-terminal state, which may already have swallowed up any retinue of planets that they once possessed.

Such inferences are based on the assumption that atoms and their nuclei are the same everywhere. Newton's great insight was to link the roles of gravity here on Earth and in celestial orbits. Yet even he only addressed the motions within our Solar System. It took much longer to realize that gravity applied in other stars, and even in other galaxies. In ancient times, the celestial sphere was believed to be made of a special substance, 'quintessence', purer than the earth, air, fire and water of our terrestrial realm. Until the mid-nineteenth century, there were no clues as to what the stars were made of. The use of prisms to disperse light into a rainbow of different colours revealed that light from the Sun, and from other stars, contained the hues characteristic of well-known atoms on Earth. The ingredients of starstuff were no different from the atoms here on the 'sub-lunary sphere'.

Astrophysicists can compute, just as easily as the Sun's evolution, the life-cycle of a star that is (say) half, twice, or ten times the mass of the Sun. Smaller stars burn their fuel more slowly. In contrast, stars ten times as heavy as the Sun – the four blue Trapezium stars in the constellation of Orion, for instance – shine thousands of times more brightly, and consume their fuel more quickly. Their lifetimes are much shorter than the Sun's, and they expire in a more violent way, by exploding as supernovae. They become, for a few weeks, as bright as several billion suns. Their outer layers, blown off at 20,000 kilometres per second, form a blast wave that ploughs into the surrounding interstellar gas.

On 24 February 1987 a Canadian astronomer, Ian Shelton, was carrying out routine observations with his Chilean assistant at the Las Campanas Observatory in northern Chile. They noticed an unfamiliar glow in the Southern sky, bright enough to be seen with the unaided eye, that had not been visible on the previous night. This proved to be much the nearest supernova to be observed in modern times. During the few weeks of its peak brilliance, and during its gradual fading in the subsequent years, it has been monitored by all the techniques of modern astronomy, allowing theories of these immense explosions to be tested. It is the only supernova whose precursor star was already known: old photographic plates show, at the site of the supernova, a blue star of about twenty solar masses.

Supernovae represent cataclysmic events in the life of the stars, involving some 'extreme' physical processes; so supernovae naturally fascinate astronomers. But only one person in ten thousand is an astronomer. What possible relevance could these stellar explosions thousands of light-years away have to all the others, whose business lies purely on or near the Earth's surface? The surprising answer is that they are fundamental to everyone's environment. Without them, we would never have existed. Supernovae have created the 'mix' of atoms that the Earth is made of and that are the building blocks for the intricate chemistry of life. Ever since Darwin, we've been aware of the evolution and selection that preceded our emergence, and of our links with the rest of the biosphere. Astronomers now trace our Earth's origins back to stars that died before the Solar System formed. These ancient stars made the atoms of which we and our planet are composed.

ALCHEMY IN THE STARS

Atoms exist in nature in ninety-two varieties, as represented in the 'periodic table'. The place of each atom in the table depends on the number of protons in its nucleus. The table starts with hydrogen at number 1, and goes up to uranium at 92. The nuclei of atoms contain not only protons but particles of another kind, called neutrons. A neutron weighs slightly more than a proton, but has no electric charge. Atoms of a particular element can exist in several variants, called isotopes, with different numbers of neutrons. For example, carbon is number 6 in the periodic table: its nucleus contains six protons. The most common form (known as ^{12}C) contains six neutrons as well; but there are also isotopes with seven and eight neutrons (known as ^{13}C and ^{14}C respectively). Uranium is the heaviest naturally occurring element; still heavier nuclei, with charges up to 114, have been made in laboratories. These super-heavy elements are unstable to fission. Some, like plutonium (number 94 in the table), have a lifetime of thousands of years; those numbered beyond 100 can be manufactured in experiments that collide nuclei together, but decay after a transient existence.

When a big star's central hydrogen has all been converted into helium (number 2), the core is pulled inwards, squeezing it hotter until the helium can itself react – helium nuclei have twice the electric charge of hydrogen, so they need to collide faster in order to overcome the fiercer electric repulsion, and this demands a higher temperature. When the helium is itself used up, the star contracts and heats up still more. Stars like the Sun never achieve hot-enough cores to permit these transmutations to go very far, but the centres of heavier stars, where gravity is more powerful, reach a billion degrees. They release further energy via the build-up of carbon (six protons), and then by a chain of transmutations into progressively heavier nuclei: oxygen, neon, sodium, silicon, etc. The amount of energy released when a particular nucleus forms depends on a competition between the nuclear force that 'glues' its constituent protons and neutrons together, and the disruptive effects of electric forces between the protons. An iron nucleus (twenty-six protons) is more tightly bound than any other; energy must be added (rather than being released) to build up still heavier nuclei. A star therefore faces an energy crisis when its core has been transmuted into iron.

The consequences are dramatic. Once the iron core gets above a threshold size (about 1.4 solar masses), gravity gains the upper hand and the core implodes down to the size of a neutron star. This releases enough energy to blow off the overlying material in a colossal explosion – creating a supernova. Moreover, this material has, by then, an 'onion skin' structure: hydrogen and helium are still burning in the outer regions, but the hotter inner layers have been processed further up the periodic table. The debris thrown back into space contains this mix of elements. Oxygen is the most common, followed by carbon, nitrogen, silicon and iron. The calculated proportions, when we take account of all types of star and the various evolutionary paths they take, agree with those observed on Earth.

Iron is only the twenty-sixth element in the periodic table, and at first sight the heavier atoms might seem a problem because it takes an *input* of energy to synthesize them. But intense heat in the collapse, and the blast wave that blows off the outer layers, together produce small traces of the elements in the rest of the periodic table, right up to uranium at number 92.

THE GALACTIC ECOSYSTEM

The first stars formed about ten billion years ago from primordial material that contained only the simplest atoms – no carbon, no oxygen, and no iron. Chemistry would then have been a very dull subject. There could certainly have been no planets around these first stars. Before our Sun even formed, several generations of heavy stars could have been through their entire life-cycles, transmuting pristine hydrogen into the basic building blocks of life and flinging them back into space via strong winds or explosions. Some of these atoms found themselves in an interstellar cloud resembling the Orion Nebula, where, about 4.5 billion years ago, a new star condensed, surrounded by a dusty disc of gas, to become our Solar System. Why are carbon and oxygen so common here on Earth, but gold and uranium so rare? The answer involves stars that exploded before our Sun formed. The Earth, and we ourselves, are the ashes from those ancient stars. Our galaxy is an ecosystem, recycling atoms again and again through generations of stars.

The carbon, oxygen and iron atoms in the Solar System are fossils from the dusty cloud from which it formed about 4.5 billion years ago: they were made by heavy stars that had already expelled processed debris by that time. These 'pollutants' constituted only two per cent of the mass: hydrogen and helium were still overwhelmingly the dominant atoms. Heavy atoms are, however, overrepresented on Earth, because hydrogen and helium are volatile gases that escaped from all the inner planets. In contrast, the giant planet Jupiter is, like the Sun, mainly hydrogen and helium. It was formed from the cooler outer part of the disc that surrounded the newly formed Sun, and its own gravity was enough to retain these lightweight atoms.

Older stars than the Sun would have formed before our galaxy had undergone so much 'pollution'; their surfaces should therefore be deficient in heavy elements compared with the Sun. Starlight has a complicated spectrum, in which each kind of atom imprints a distinctive set of colours. (Streetlights have, for instance, familiarized us with the yellow light of sodium, and the characteristic blue of mercury vapour.) All the heavier atoms, indeed, tend to be less abundant in the oldest stars, corroborating this general scheme of galactic history. Helium, in contrast, is very abundant even in the oldest stars, and the reason, discussed in the next chapter, leads us back to the first few minutes after the Big Bang.

NUCLEAR EFFICIENCY: ε = 0.007

Accounting for the proportions of the different atoms – and realizing that the Creator didn't need to turn ninety-two different knobs – is a triumph of astrophysics. Some details are still uncertain, but the essence depends on just one number: the strength of the force that binds together the particles (protons and neutrons) that make up an atomic nucleus.

Einstein's famous equation $E = mc^2$ tells us that mass (m) is related to energy (E) via the speed of light(c). The speed of light thus has fundamental significance. It fixes the 'conversion factor': it tells us how much each kilogram of matter is 'worth' in terms of energy. The only way that mass can be converted 100 per cent into energy is if it can be brought together with an equal mass of antimatter – something that (fortunately for our survival) doesn't exist in bulk anywhere in our galaxy. Just one kilogram of antimatter would yield as much energy as a large electrical power station generates in ten years. But ordinary fuels such as gasoline, or even explosives such as TNT, release only about a *billionth* of the material's 'rest mass energy'. Such materials involve chemical reactions, which leave the nuclei of atoms unchanged and just reshuffle the orbits of their electrons and the linkages between the atoms. But the power of nuclear fusion is awesome because it is millions of times more efficient than any chemical explosion. The nucleus of a helium atom weighs 99.3 per cent as much as the two protons and two neutrons that go to make it. The remaining 0.7 per cent is released mainly as heat. So the fuel that powers the Sun – the hydrogen gas in its core – converts 0.007 of its mass into energy when it fuses into helium. It is essentially this number, **ε**, that determines how long stars can live. Further transmutations of helium all the way up to iron release only a further 0.001. The later stages in a star's life are therefore relatively brief. (They are even briefer because, in the hottest stellar cores, extra energy drains away invisibly in neutrinos.)

The amount of energy released when simple atoms undergo nuclear fusion depends on the strength of the force that 'glues' together the ingredients in an atomic nucleus. This force is different from the two forces I have discussed so far, namely gravity and electricity, because it acts only at very short range, and is only effective on the scale of an atomic nucleus. We don't directly experience it, in contrast to the way that we can 'feel' electrical and gravitational forces. Within an atomic nucleus, however, this force grips the protons and neutrons together strongly enough to combat the electrical repulsion that would otherwise make the (positively charged) protons fly apart. Physicists call this force the 'strong interaction'.

This 'strong' force, the dominant force in the microworld, holds the protons in helium and heavier nuclei together so firmly that fusion is a powerful enough energy source to provide the prolonged warmth from the Sun that was a prerequisite for our emergence. Without nuclear energy, the sun would deflate within about ten million years, as Kelvin realized a century ago. Because the force only acts at short range, it becomes less effective in the larger and heavier nuclei: this is why the nuclei heavier than iron become less tightly bound rather than more so.

THE TUNING OF ε

Nuclear forces are crucial, but how much does their exact strength matter? What would change if ε were, for instance, 0.006 or 0.008 rather than 0.007? At first sight, one might suspect that it wouldn't make much difference. If ε were smaller, hydrogen would be a less efficient fuel and the Sun and stars wouldn't live so long, but this in itself would not be crucial – after all, we are here already, and the Sun is still less than halfway through its life. But there turn out to be delicate effects, sensitive to this number, in the synthesis process that transforms hydrogen into the rest of the periodic table.

The crucial first link in the chain – the build-up of helium from hydrogen – depends rather sensitively on the strength of the nuclear 'strong interaction' force. A helium nucleus contains two protons, but it also contains two neutrons. Rather than the four particles being assembled in one go, a helium nucleus is built up in stages, via deuterium (heavy hydrogen), which comprises a proton plus a neutron. If the nuclear 'glue' were weaker, so that Ɛ were 0.006 rather than 0.007, a proton could not be bonded to a neutron and deuterium would not be stable. Then the path to helium formation would be closed off. We would have a simple universe composed of hydrogen, whose atom consists of one proton orbited by a single electron, and no chemistry. Stars could still form in such a universe (if everything else were kept unchanged) but they would have no nuclear fuel. They would deflate and cool, ending up as dead remnants. There would be no explosions to spray the debris back into space so that new stars could form from it, and no elements would exist that could ever form rocky planets.

At first sight, one might have guessed from this reasoning that an even stronger nuclear force would have been advantageous for life, by making nuclear fusion more efficient. But we couldn't have existed if Ɛ had been more than 0.008, because no hydrogen would have survived from the Big Bang. In our actual universe, two protons repel each other so strongly that the nuclear 'strong interaction' force can't bind them together without the aid of one or two neutrons (which add to the nuclear 'glue', but, being uncharged, exert no extra electrical repulsion). If Ɛ were to have been 0.008, then two protons would have been able to bind directly together. This would have happened readily in the early universe, so that no hydrogen would remain to provide the fuel in ordinary stars, and water could never have existed.

So any universe with complex chemistry requires Ɛ to be in the range 0.006–0.008. Some specific details are still more sensitive. The English theorist Fred Hoyle stumbled on the most famous instance of 'fine tuning' when he was calculating exactly how carbon and oxygen were synthesized in stars. Carbon (with six protons and six neutrons in its nucleus) is made by combining three helium nuclei. There is negligible chance of all three coming together simultaneously, and so the process happens via an intermediate stage where two helium nuclei combine into beryllium (four protons and four neutrons) before combining with another helium nucleus to form carbon. Hoyle confronted the problem that this beryllium nucleus is unstable: it would decay so quickly that there seemed little chance of a third helium nucleus coming along and sticking to it before it decayed. So how could carbon ever arise? It turned out that a special feature of the carbon nucleus, namely the presence of a 'resonance' with a very particular energy, enhances the chance that beryllium will grab another helium nucleus in the brief interval before it decays. Hoyle actually predicted that this resonance would exist; he urged his experimental colleagues to measure it, and was vindicated. This seeming 'accident' of nuclear physics allows carbon to be built up, but no similar effect enhances the next stage in the process, whereby carbon captures another helium nucleus and turns into oxygen. The crucial 'resonance' is very sensitive to the nuclear force. Even a shift by four per cent would severely deplete the amount of carbon that could be made. Hoyle therefore argued that our existence would have been jeopardized by even a few percentage points' change in ε.[1]

Irrespective of how the elements were made, a change in ε would affect the length of the periodic table. A weaker nuclear force would shift the most tightly bound nucleus (which is now iron, number 26) lower down the periodic table and reduce below ninety-two the number of stable atoms. This would lead to an impoverished chemistry. Conversely, a larger ε could enhance the stability of heavy atoms.

At first sight, a longer 'menu' of different abundant atoms would seem to open the way to a more interesting and varied chemistry. But this isn't altogether obvious – for example, the English language would not be enriched in any important sense if the alphabet had more letters in it. Likewise, complex molecules can exist in endless variety even though there are relatively few common elements. Chemistry would be duller (and complex molecules of the kind essential for life would not exist) if there were no oxygen and iron (numbers 8 and 26 respectively), and especially if carbon (number 6) were not abundant; but little would be added by enhancing the number of abundant elements, or by having a few extra stable elements beyond our natural ninety-two.

The actual mix of elements would depend on Ɛ, but what is remarkable is that no carbon-based biosphere could exist if this number had been 0.006 or 0.008 rather than 0.007.

CHAPTER 5

OUR COSMIC HABITAT II: BEYOND OUR GALAXY

> *Telescope* (n): A device having a relation to the eye similar to that of a telephone to the ear, enabling distant objects to plague us with a multitude of needless details.
>
> Ambrose Bierce

THE UNIVERSE OF GALAXIES

I've described how the atoms of the periodic table are made: that we're stardust – or, less romantically, the 'nuclear waste' – from the fuel that makes stars shine. These processes depend on the strength of the 'nuclear force' that glues together the protons and neutrons within the nuclei of these atoms – measured by the cosmic number ε = 0.007 that denotes the proportion of energy that is released when hydrogen fuses into helium. But where did the original protons and hydrogen atoms come from, and how did the primordial material aggregate into the first galaxies and stars? To answer these questions, we must extend our horizons in space and time – out to the extragalactic realm, and back to an era before the birth of the first stars. We shall encounter further numbers that describe our entire universe, and discover that our emergence depended on these too being finely tuned.

Stars are agglomerated into galaxies, which are the basic

units that make up the universe. Our own is typical. Its hundred billion stars lie mainly in a disc, circling around a bright inner 'bulge' where the stars are closer together than average. Right at the centre lurks a black hole with the mass of 2.5 million suns. A light signal would take about 25,000 years to reach us from the galactic centre, and we on Earth are rather more than halfway out towards the disc's edge. From our Sun's location, the other stars in the disc appear concentrated in a band across the sky, known to us as Milky Way. Typical stars take more than a hundred million years for a single circuit (sometimes called a 'galactic year') around the galactic centre.

Andromeda, our galaxy's nearest major neighbour in space, is about two million light-years away. To an astronomer on a planet orbiting one of Andromeda's stars, our galaxy would look rather like Andromeda does to us: a disc, viewed obliquely, made of stars and gas circling around a central 'hub'. Millions of other galaxies are visible with large telescopes. Not all are disc-like: the other important class is the so-called 'elliptical galaxies', in which the stars are not organized into a disc but are swarming around in more random orbits, each feeling the gravitational pull of all the others.

Galaxies are not sprinkled around randomly in space: most are in groups or clusters, held together by gravity. Our own Local Group, a few million light-years across, contains the Milky Way and Andromeda, together with thirty-four smaller galaxies (that, at least, was the last count – very faint and small members of the Local Group are still being discovered). Gravity is pulling Andromeda towards us at about 100 kilometres per second. In about five billion years, these two disc galaxies may crash together. Such crashes are 'routine' cosmic events: we see, deeper in space, many other galaxies that seem to be undergoing such an encounter with another.

Galaxies are so vast and diffuse, and the stars are so thinly spread, that actual collisions between individual stars are exceedingly rare. (This is clearly true in the Solar neighbourhood, because even the nearest stars seem like faint points of light). Even when two galaxies crash together and merge, there would be very few stellar impacts. All that happens is that each star feels the collective gravity of everything in the other galaxy. Orbits are so distorted that the stars end up in a single chaotic swarm rather than two separate discs. This is, of course, just what a so-called elliptical galaxy looks like, and I suspect (though the issue is still controversial) that the big elliptical galaxies were formed in this fashion.

THE TEXTURE OF OUR UNIVERSE: THE COSMIC WEB

Our Local Group is near the edge of the Virgo Cluster, an archipelago of several hundred galaxies, whose core lies about fifty million light-years away. The clusters and groups are themselves organized into still larger aggregates. The so-called 'Great Wall', a sheet-like array of galaxies about 200 million light-years away, is the nearest and most prominent of these giant features. Another concentration of mass, the 'Great Attractor', exerts a gravitational force that pulls us, and the entire Virgo Cluster as well, at several hundred kilometres per second.

Many phenomena in nature – mountain landscapes, coastlines, trees, blood vessels, and so forth – are 'fractals'. A fractal is a pattern with the special mathematical feature that a small part, when magnified, resembles the whole. If our universe were like this – if it contained clusters of clusters of clusters . . . *ad infinitum* – then however deeply we probed into space, and however large a volume we sampled, the galaxies would still have a patchy distribution: by probing deeper, we'd simply be sampling larger and larger scales in the clustering hierarchy. But this is not how our universe looks. Powerful telescopes reveal galaxies out to several billion light-years. Within this far larger volume, astronomers have mapped many more clusters like Virgo, and more features like the 'Great Wall'. But deeper surveys don't reveal any conspicuous features on still larger scales; in the words of the Harvard astronomer Robert Kirshner, we reach 'the end of greatness'. A box whose sides are 200 million light-years (a distance still small compared with the horizon of our observations, which is about 10 billion light-years away) is capacious enough to accommodate the largest structures, and to contain a 'fair sample' of our universe. Wherever it is placed, such a box would contain roughly the same number of galaxies, grouped in a statistically similar way into clusters, filamentary structures, etc. The hierarchy of clustering doesn't continue towards indefinitely large scales.

Our universe is thus not a simple fractal; moreover the 'smoothing scale' is small compared with the largest distances that our telescopes can probe. As an analogy, imagine you were on a ship in the middle of the ocean. A complicated pattern of waves would surround you, stretching to the horizon. But you could study the statistics of the waves because your field of view extends far enough to encompass many of them. Even the biggest waves on the ocean are far smaller than the horizon distance, and you could, in your imagination, divide what you can see into many separate patches, each large enough to be a fair sample. There is a contrast here between seascapes and landscapes: in mountainous terrain, one grand peak often dominates the entire horizon and you can't define meaningful averages as you can for a seascape. (Landscapes, indeed, *can* be fractal-like. The mathematics of fractals is used in computer graphics programs for depicting imaginary landscapes in movies.)

Cosmic structures encompass a wide range of dimensions: stars, galaxies, clusters, and superclusters. On scales less than about 1/300 of the horizon, the concentration of galaxies varies by more than a factor of two from place to place; on larger scales, the fluctuations are gentler (though there are a few conspicuous features like the Great Attractor). Superclusters of galaxies – to extend the ocean analogy – are like the longest conspicuous waves. We shall see in Chapter 8 that this scale depends on a single cosmic number, *Q*, imprinted in the very early universe, and that the 'embryos' of clusters and superclusters – structures stretching millions of light-years across the sky – can be traced back to a time when the entire universe was of microscopic size. This is perhaps the most astonishing link between the outer space of the cosmos and the inner space of the microworld.

One's first guess might be that the texture of our universe on such large scales was irrelevant to our local habitat within the Solar System: it might not seem to matter whether our galaxy contained a quadrillion stars, or else 'only' a million, rather than the hundred billion that we observe; nor whether it belonged to a cluster containing millions of other galaxies rather than just a few. But a universe much rougher than ours wouldn't be hospitable to stars and planets. On the other hand, a universe that was too smooth would be blandly uninteresting: no galaxies and stars would form, and all the material would be thinly spread and amorphous.

This will be the theme of Chapter 8. But, for the moment, we can note another crucial consequence of the large-scale smoothness: it makes the science of cosmology possible, by allowing us to define the average properties of our universe – the demography of the galaxies, the statistics of the clusters, and so forth. Despite galaxies and clusters, it is still useful to think about the smoothed-out properties of the universe, just as we can describe the Earth as 'round', despite the complex topography of its mountains and its ocean depths. However, it would not be useful to describe the Earth as 'basically round' if its mountains were thousands of kilometres high rather than just a few.

Even more important, we can meaningfully ask whether our entire universe is static, expanding or contracting.

THE EXPANSION

Galaxies are the 'building blocks' of our universe, and by studying the light from them we can infer how they are moving. The hundred billion stars in a typical galaxy are too faint to be seen individually: telescopes record the total light from many stars blurred together. This light can be analysed into a spectrum. We have noted how the light from a single star can reveal its speed towards (or away from) us, and how repeated measurements can even pick up the tiny oscillatory motion induced by an orbiting planet. Likewise, the spectrum of an entire galaxy reveals how fast it is moving, either towards us (a shift towards the blue end of the spectrum) or away from us (a shift towards the red).

Perhaps the most important single fact about our universe is that the light from all distant galaxies is shifted towards the red: all (except for a few nearby galaxies in the same cluster as our own) are receding from us. Moreover, the redshift (a measure of the recession speed) is larger for the more distant galaxies. We seem to be in an expanding universe, where clusters of galaxies get more widely separated – more thinly spread through space – as time goes on.

The simple relation between redshift and distance is named after Edwin Hubble, who first claimed such a law in 1929. Observers on other galaxies would witness a similar expansion of distant regions away from them. The expansion is a *broad-brush* effect: individual galaxies (even clusters of galaxies) are not themselves expanding; still less does the expansion affect anything more local, such as our Solar System.

Imagine that the rods in the M. C. Escher drawing in Figure 5.1 lengthened at the same rate. An observer on any vertex would see the others receding at speeds that depended on how many intervening rods there were. In other words, the recession speed of other vertices would be proportional to their distance. The galaxies aren't in a regular lattice – as

FIGURE 5.1
Escher's *Cubic Space Division*. If the rods in this lattice all lengthened at the same rate, the vertices would recede from each other in accordance with Hubble's law; but no vertex is 'special' and there is no centre.

already mentioned, they are in groups or clusters – but you can nonetheless envisage the expansion by imagining that clusters of galaxies are linked by rods that all lengthen at the same rate. There is nothing special about any vertex in the picture; and there is likewise nothing special about the location of our galaxy in the universe. (Although our galaxy is randomly *placed*, we are not, however, observing it at a random *time*; the reasons for this will become clear later.) Cosmology has only progressed because our universe, in its large-scale structure, is uniform enough to be described by a simple 'Hubble expansion', where all patches seem to be expanding similarly.

The expansion can be envisaged locally as a Doppler effect, but on large scales, when the apparent recession is at a good fraction of the speed of light, it is better to attribute the redshift to a 'stretching' of space while the light travels through it. The amount of reddening – in other words, the amount that the wavelengths are stretched – is then equal to the amount by which the universe has expanded (and, in our Escher analogy, the 'rods' have lengthened) while the light has been travelling towards us.

We might of course wonder whether the redshift actually implies expansion, rather than some new physical effect that comes into play over long distances. The possibility of such a 'tired light' effect is still sometimes raised, although nobody has come up with a viable theory consistent with all the evidence (it must, for instance, produce the same fractional change in wavelength for light of all colours, and mustn't blur the images of distant objects). A non-expanding universe would actually entail even worse paradoxes than any Big Bang theory. Stars don't have infinite energy reserves; they evolve, and eventually exhaust their fuel. So therefore do galaxies, which are essentially aggregates of stars. It is possible to date the oldest stars in our Milky Way, and in other galaxies, by comparing their properties with the outcome of computations of how stars evolve. The oldest are about ten billion years old – entirely consistent with the view that our universe has only been expanding for a bit longer than that. If our universe were static, all galaxies must have mysteriously 'switched on' in their present positions – in a synchronized fashion – about ten billion years ago. A non-expanding universe would entail severe conceptual difficulties.

The expansion almost certainly began between ten and fifteen billion years ago, twelve or thirteen billion being the best guess. There are two reasons for this persistent uncertainty in the age of our universe. The exact distances to galaxies are (unlike their recession speeds) still somewhat inexact; also, the estimate depends on how much faster (or slower) the expansion might have been in the past.

SEEING INTO THE PAST

Light travels at a finite speed, and so we see distant regions not as they are now but as they were a long time ago. At earlier epochs, the universe would have been more compressed – the rods in our lattice would have been shorter. So the second Escher picture, *Angels and Devils*, shown in Figure 5.2, better represents what we actually see.

We'd expect very distant galaxies to look different from nearby ones. Their light has taken a long time on its journey, and so they were younger and less evolved when they emitted

FIGURE 5.2

Escher's *Angels and Devils*. Because of light's finite speed, we see remote regions as they were in the remote past. Towards the horizon, everything appears more compressed.

the light now reaching us. Not all the pristine gas had at that stage condensed into stars. These evolutionary changes would be so slow that they would only be manifest over billions of years. To detect a trend, one must therefore probe galaxies so far away that their light set out several billion years ago.

The Hubble Space Telescope (HST) – named in honour of the discoverer of cosmic expansion – circles the Earth far above the blurring effect of the atmosphere and has produced the sharpest pictures yet of very distant regions. The HST is so sensitive that a long exposure reveals, close packed in the sky, literally hundreds of faint smudges, even within a field of view so small that it would cover less than a hundredth of the area of the full moon – and would appear as a blank patch of sky when viewed with an ordinary telescope. (I think that the amazing pictures being generated by the HST will impact as strongly on public consciousness as the first images from space, in the 1960s, that showed the whole Earth, with its delicate-seeming biosphere.) The faint features in these pictures, with a diversity of shapes, are a billion times fainter than any star we can see with the unaided eye. But each is an entire galaxy, thousands of light-years in size, which appears so small and faint because of its huge distance from us. These galaxies look different from their nearby counterparts because they are being viewed when they have only recently formed: they have not yet settled down into steadily spinning discs like the photogenic nearby spiral galaxies depicted in most astronomy books. Some consist mainly of glowing diffuse gas, which hasn't yet fragmented into stars. Most of them appear intrinsically bluer than present-day galaxies (after correcting, of course, for the redshift), because massive blue stars, which would all by now have died, were still shining when the light left these distant galaxies.

These very deep images show us what a galaxy like our own Milky Way would have looked like when its first stars were shining brightly. When we observe Andromeda, a nearby 'twin' of our own galaxy, we may wonder whether Andromedans are looking back at us with still bigger telescopes.

Perhaps they are. But there could be nothing so 'advanced' on these very remote galaxies: we are viewing them at a very primitive evolutionary stage, before enough time has elapsed for many stars to have completed their lives. They have as yet no complex chemistry; there is very little oxygen, carbon, etc, even to make planets; and so there is scant chance of life. We are seeing these galaxies at a stage when the basic building blocks for planetary systems were being laid down. (The light we detect was actually emitted in the far ultra-violet. Such radiation cannot be detected by the eye, nor can it even penetrate the Earth's atmosphere. But the extreme ultra-violet radiation from these galaxies, by the time it reaches us, has been shifted into red light.)

The most distant galaxies are so redshifted that the wavelength of the light has been stretched by more than a factor of six: that's how much the universe must have expanded since the light set out. If the expansion had been steady, with galaxies neither accelerating nor decelerating, then when the universe was one-sixth its present scale (in the sense that distances – the rods on Escher's lattice – were scaled down six times smaller) it would have been one-sixth its present age. This statement might at first seem troublesome: doesn't it mean that a galaxy must be moving away at five times the speed of light, if the light has taken five-sixths of our universe's present age to get back to us? But there's no paradox. Einstein's special theory of relativity tells us that nothing can move faster than light, relative to us, *when time is measured by our clock*. But that theory also tells us that a fast-moving clock runs slow. A fast clock can indeed travel five light-years for every year that it records if it moves at ninety-eight per cent of the speed of light.

The situation is actually a bit more complicated because the recession speed would not be constant. The gravitational pull that everything in the universe exerts on everything else causes deceleration, which tends to make the earlier stages of cosmic expansion relatively even shorter. But (as discussed in Chapter 7) another force may be at work that tends to speed up the expansion. There is still, therefore, some uncertainty about how far back in time (or how far away in space) these remote galaxies actually are: the best guess would be that their light set out when the universe was around one-tenth of its present age.

Cosmologists study 'fossils' of the past: old stars, chemical elements synthesized when our galaxy was young, etc. In that respect they are like geologists or palaeontologists trying to infer how our Earth and its fauna have evolved. But cosmologists actually have an edge over other scientists who can't do experiments and depend on 'historical' evidence. By directing their telescopes towards distant objects, cosmologists can *see* the evolution they claim: populations of distant galaxies, whose light set out several billion years ago, look different from their counterparts nearby. Because of the large-scale uniformity, all parts of the universe have had similar histories. These remote galaxies should therefore – statistically at least – look similar to the way our Milky Way, the Andromeda galaxy and other nearby systems would have looked billions of years ago.

The field of view of a telescope is a long thin cone, extending out to the limits of vision. The objects at each distance tell us about a specific epoch in the past. As we probe greater distances, we probe further back in time, just as a borehole through successive layers of Antarctic ice can reveal the history of the Earth's climate.

The Hubble Space Telescope was dogged by delays, errors and cost overruns, but it has now, albeit belatedly, fulfilled the hopes that astronomers had for it. Its out-of-focus mirror was corrected by the first manned repair mission in 1994; and the light detectors on board have been upgraded. It could, barring mishap, continue until 2010, by which time still larger space telescopes may have been deployed. But equally important has been the advent of a new generation of larger telescopes on the ground. Their 8–10 metre mirrors offer sixteen times more collecting area than the HST, and so they can collect far more light from very faint and remote galaxies.

The two Keck Telescopes, on Mauna Kea in Hawaii, were the first of these new-generation instruments to come into service, but there are now several more. Most impressive of all is the Very Large Telescope (VLT), a connected cluster of four telescopes, each with an eight-metre mirror, constructed in the Chilean Andes by a consortium of European nations.

The sharpness of the images from these ground-based telescopes is limited by the blurring due to turbulence in the atmosphere (the same process that makes stars twinkle). This limit can be surmounted either by linking two or more telescopes together and combining the images, or by so-called 'adaptive optics', whereby a mirror is continually tweaked and adjusted to compensate for fluctuations in the atmosphere.

These superb instruments offer snapshots of the universe right back to when the first galaxies were forming. The first *stars* may actually have formed even earlier, in aggregates smaller than present-day galaxies, but which are too faint for us to see. These later agglomerated into larger structures. The rate at which gas condenses into stars is the 'metabolic rate' of a galaxy. It seems to have peaked when the universe was about a quarter of its present age (even though the very first starlight appeared much earlier). Fewer bright stars are forming now because most of the gas in 'mature' galaxies has already been incorporated into older stars.

That, at least, is the scenario that most cosmologists accept. Fleshing out the details will need more observations and a fuller understanding of how stars form. The aim is to obtain a consistent scenario that not only matches all we know about present-day galaxies but also takes into account the increasingly detailed snapshots of what they looked like, and how they were clustered, at all earlier times. When data are sparse, they may all fit with several completely wrong theories; but as the evidence mounts up, we should 'home in' on a single picture of how things work.

With increasing distance our knowledge fades and fades rapidly. Eventually we reach the dim boundary, the utmost

> limits of our telescope. There we measure shadows, and we search among ghostly errors of measurement for landmarks that are scarcely more substantial. The search will continue. Not until the empirical resources are exhausted need we pass on to the dreamy realm of speculation.

These are the concluding words of Edwin Hubble's classic (1936) book, *The Realm of the Nebulae.* Recent progress would have delighted, and probably astonished, Hubble. That progress is owed to the telescope in space that bears his name, and huge new telescopes on the ground.

BEFORE THE GALAXIES

What about still earlier epochs, before any galaxies could have formed? The best evidence that everything really emerged from a dense 'beginning' is that intergalactic space isn't completely cold. This warmth is an 'afterglow of creation'. It manifests itself as microwaves, the kind of radiation that generates heat in a microwave oven but very much less intense. The first detection of the 'cosmic microwave background', back in 1965, was the most important advance in cosmology since the discovery of the expansion of the universe. Later measurements confirmed that these microwaves have a very distinctive property: their intensity at different wavelengths, when plotted on a graph, traces out what physicists call a 'black body' or 'thermal' curve. This particular curve is expected when the radiation has been brought into balance with its environment (as happens deep inside a star, or in a furnace that has burnt steadily for a long time); it's just what would be expected if the microwaves were indeed a relic of a 'fireball' phase when everything in our universe was squeezed hot, dense and opaque.

By far the most precise measurements came during the 1990s from NASA's Cosmic Background Explorer Satellite

(COBE). When experimenters present their results, they conventionally draw 'error bars' indicating the range of uncertainty, but for the COBE data the 'bars' can't be exhibited because they would be shorter than the thickness of the curve. This truly remarkable measurement, with an accuracy of one part in 10,000, confirms beyond reasonable doubt that everything in our universe – all the stuff that galaxies are now made of – was once a compressed gas, hotter than the Sun's core.

The present average temperature of the universe is 2.728 degrees above absolute zero. This is, of course, exceedingly cool (around −270°C); but there's a well-defined sense in which intergalactic space still contains a lot of heat. Every cubic metre contains 412 million quanta of radiation, or photons: in comparison, the average density of *atoms* in the universe is only about 0.2 per cubic metre. This latter number is less precisely known, because we are unsure how many atoms may be in diffuse gas or 'dark' matter, but there seem to be about two billion photons for every atom in the universe. During the expansion of the universe, the density of atoms and of photons both decrease. But the decrease is the same for both, and so the ratio of photons to atoms stays the same. Because this ratio of 'heat' to 'matter' is so large, the early universe is often referred to as a 'hot' Big Bang.

The hot early phases wouldn't have lasted long. Only for a few minutes would the temperature have exceeded a billion degrees. After about half a million years it had cooled to 3000 degrees – a bit cooler than the Sun's surface. This marks a significant stage in the expansion process: before that time, everything was so hot that electrons were dislodged from nuclei and moved freely; but afterwards the electrons would have slowed down enough to attach themselves to nuclei, forming neutral atoms. These atoms cannot scatter the radiation as efficiently as free electrons were able to during the earlier and hotter stages. The primordial material would thereafter have been transparent; the 'fog' would have lifted. During expansion, the temperature drops inversely with the scale of the universe (the length of the rods on Escher's lattice). The microwaves that COBE detects are relics from the era when our universe was more than a thousand times more compressed – at 3000 degrees rather than 2.7 degrees, and long before any galaxies came into existence. The intense radiation in the original fireball, although cooled and diluted by expansion, still pervades the whole universe.

The often-used analogy with an explosion is misleading inasmuch as it conveys the image that the Big Bang was triggered at some particular centre. But as far as we can tell, any observer – whether on Earth, on Andromeda, or even on the galaxies remotest from us – would see the same pattern of expansion. The universe may once have been squeezed to a single point, but everyone had an equal claim to have started from that point; we can't identify the origin of the expansion with any particular location in our present universe.

It is also incorrect to think of the *high pressure* in the early universe 'driving' the expansion. Explosions are caused by an *unbalanced* pressure; bombs on Earth, or supernovae in the cosmos, explode because a sudden boost in internal pressure flings debris into a low-pressure environment. But in the early universe the pressure was the same everywhere: there was no edge, no 'empty' region outside. The primordial gas cools and dilutes, just as happens to the contents of an expanding box. The extra gravity due to the pressure and heat energy actually slows down the expansion.[1]

This is a consistent picture, but it leaves some mysteries. Above all (since the explosion analogy is flawed) it offers no explanation for *why* expansion occurs at all. The standard Big Bang theory simply postulates that everything was set up with just enough energy to go on expanding. An answer to why it is expanding at all must be sought in the still earlier stages, where we don't have such direct evidence nor such a confident understanding of the physics.

The name 'Big Bang' was introduced in the 1950s by the celebrated Cambridge theorist Fred Hoyle (already mentioned in Chapter 4 for his insights into the origin of carbon) as a derisive description of a theory he didn't like. Hoyle himself favoured a 'steady state' universe, in which new atoms and new galaxies were imagined to form continuously in the gaps as the universe expanded, so that its average properties never changed. There was at that time no evidence either way – cosmology was the province of armchair speculators – because observations didn't probe far enough for the evolution (if it existed) to show up. But the steady-state theory fell from favour as soon as evidence emerged that the universe was actually different in the past. Though it turned out wrong, the steady-state theory was a 'good' theory in that it made very clear-cut and testable predictions; it was a genuine stimulus to the subject, goading observers to push their techniques to the limit. (A 'bad' theory, in this sense, is one that is so flexible that it can be adjusted to account for any data. The eminent – and arrogant – physicist Wolfgang Pauli would deride such vague ideas as 'not even wrong'.) Hoyle himself never became fully reconciled to the Big Bang, although he adopted a compromise picture that sceptical colleagues called a 'Steady Bang'.

NUCLEAR REACTIONS IN THE BIG BANG

According to the Big Bang theory, our universe started off hotter than the centre of a star. Why, then, weren't the primordial nuclei of hydrogen all transmuted into iron during the Big Bang? (Remember that nuclei of iron are more 'tightly bound' than any others, and are built up in the cores of the biggest and hottest stars.) If this had happened, no long-lived stars could have existed in our present universe, because all the available fuel would have been used up in the early fireball: a star made of vaporized iron could exist, but it would deflate within millions of years, instead of billions, rather as Kelvin thought the Sun would. Fortunately, the first few minutes of the expansion didn't allow enough time for nuclear reactions to 'process' any of the primordial material into iron – nor even into carbon, oxygen, etc. The reactions would turn about twenty-three per cent of the hydrogen into helium, but (apart from a trace of lithium) no elements higher up the periodic table emerge from the Big Bang itself.

This primordial helium is, however, crucial and offers us strong corroboration of the Big Bang theory. Even the oldest objects (in which pollution by carbon, oxygen and so forth is a hundred times less than in the Sun) turn out to contain 23–24 per cent of helium: no star, galaxy or nebula has been found where helium is less abundant than this. It seems as though the galaxy started not as pure hydrogen, but was already a mix of hydrogen and helium. (The Sun's outer layers have twenty-seven per cent helium, the extra 3–4 per cent being just about what would have been made, along with carbon, oxygen and iron, in the short-lived early stars that must already have polluted the cloud from which our Solar System formed.)[2]

Many slow-burning low-mass stars survive, which formed several billion years before our Sun when our galaxy was young. These contain far less carbon, oxygen and iron relative to hydrogen than the Sun does – something that is, of course, natural if, as Hoyle was first to argue, these atoms were expelled from massive stars and accumulated gradually over galactic history. Hoyle's view contrasted with George Gamow's idea that the entire periodic table was 'cooked' in the early universe. If Gamow had been right and these elements predated the first stars and galaxies, their abundance would be the same everywhere, in young and old stars alike.

Helium is the only element that, according to calculations, would be created prolifically in a Big Bang. This is gratifying because it explains why there is so much helium and why the helium is so uniform in its abundance. Attributing helium to the Big Bang thus solved a long-standing problem, and emboldened cosmologists to take the first few seconds of cosmic history seriously.

As a bonus, the Big Bang accounts for another kind of atom: deuterium (also known as 'heavy hydrogen'). An atom of deuterium contains not just a proton but a neutron as well, which adds extra mass but no extra charge. The existence of deuterium is otherwise a mystery, because it is destroyed rather than created in stars: as a nuclear fuel it is easier to ignite than ordinary hydrogen, and so newly-formed stars would burn up any deuterium during their initial contraction, before settling down in their long hydrogen-burning phases.

Helium and deuterium were made when the temperature in the compressed universe was (in round numbers) three billion degrees – about a billion times higher than it is now. As the universe expands, we can imagine the rods of Escher's lattice (see Figure 5.1) lengthening. The wavelengths of the radiation stretch in proportion to the length of the rods, and the temperature decreases as the inverse of the length. This means that, when the temperature was around three billion degrees (rather than around 3 degrees as it is now) the rods were a billion times (10^9) shorter and the densities higher by the cube of that factor, 10^{27}. But our present universe is so diffuse – around 0.2 atoms per cubic metre – that even when compressed by this huge factor, the density is still less than that of air! The temperature was then so high that the individual nuclei would have been in agitated rapid motion. Laboratory experimenters can check what happens when hydrogen and helium nuclei crash together with the same energies as they would have had when the helium-formation occurred, so the calculations are based on quite conventional and firmly-based physics.

If we assume a present density of 0.2 atoms per cubic metre, the computed proportions of hydrogen, helium and deuterium that would emerge from the cooling 'fireball' universe agree with observations. This is gratifying, because the observed abundances could have been entirely out of line with the predictions of any Big Bang; or they might have been consistent, but only for a density that was far below, or else far above, the range allowed by observation. As we have seen, 0.2 atoms per cubic metre is indeed close to the smoothed out density of galaxies and gas in our universe. (This has important implications for 'dark matter', as discussed in the next chapter.)

CHAPTER 6

THE FINE-TUNED EXPANSION: DARK MATTER AND Ω

Eternity is very long, especially towards the end.
Woody Allen

THE CRITICAL DENSITY

In about five billion years the Sun will die; and the Earth with it. At about the same time (give or take a billion years) the Andromeda galaxy, our nearest big galactic neighbour, which belongs to the same cluster as our galaxy and which is actually falling towards us, will crash into the Milky Way.

These gross long-range forecasts are reliable because they depend on assuming that basic physics within the Sun, and the force of gravity in stars and galaxies, operate during the next five billion years as they have for the last five to ten billion. Not much of the (more interesting) detail is predictable, however. We can't be sure that the Earth will still be the third-closest planet to the Sun throughout the next five billion years: even planetary orbits can behave 'chaotically' over that expanse of time. And of course the changes on the Earth's surface, particularly the ever-more-rapid alterations in its biosphere being wrought by our own species, can't be confidently predicted even for a *millionth* of that timespan.

The Sun hasn't even burnt up half its fuel yet. More time lies ahead of it than has elapsed in the entire course of biological evolution. And the galaxy will far outlast the Sun. Even if life were now unique to Earth, there would be abundant time for it to spread through the galaxy and beyond. Manifestations of life and intelligence could eventually affect stars or even galaxies. I forbear to speculate further, not because this line of thought is intrinsically absurd but because it opens up such a variety of conceivable scenarios – many familiar from science fiction – that we can predict nothing. In contrast, long-range forecasts for our entire universe are on surer ground.

Our galaxy will surely end five or six billion years hence in a great crash. But will our universe go on expanding for ever? Will the distant galaxies move ever further away from us? Or could these motions eventually reverse, so that the entire firmament eventually recollapses to a 'Big Crunch'?

The answer depends on the 'competition' between gravity and the expansion energy. Imagine that a large asteroid or a planet were to be shattered into fragments. If the fragments dispersed rapidly enough, they would fly apart for ever. But if the disruption were less violent, gravity might reverse the motions, so that the pieces fell back together again. It's similar for any large domain within our universe: we know the expansion speed now, but will gravity bring it to a halt? The answer depends on how much stuff is exerting a gravitational pull. The universe will recollapse – gravity eventually defeating the expansion, unless some other force intervenes – if the density exceeds a definite critical value.

We can readily calculate what this critical density is. It amounts to about five atoms in each cubic metre. That doesn't seem much; indeed, it is far closer to a perfect vacuum than experimenters on Earth could ever achieve. But the universe actually seems to be emptier still.[1]

Suppose our star, the Sun, were modelled by an orange. The Earth would then be a millimetre-sized grain twenty metres away, orbiting around it. Depicted to the same scale, the nearest stars would be 10,000 kilometres away: that's how thinly spread the matter is in a galaxy like ours. But galaxies are, of course, especially high concentrations of stars. If all the stars from all the galaxies were dispersed through intergalactic space, then each star would be several hundred times further from its nearest neighbour than it actually is within a typical galaxy – in our scale model, each orange would then be *millions* of kilometres from its nearest neighbours.

If all the stars were dismantled and their atoms spread uniformly through our universe, we'd end up with just one atom in every ten cubic metres. There is about as much again (but seemingly no more) in the form of diffuse gas between the galaxies. That's a total of 0.2 atoms per cubic metre, twenty-five times less than the critical density of five atoms per cubic metre that would be needed for gravity to bring cosmic expansion to a halt.

HOW MUCH DARK MATTER?

The ratio of the actual density to the critical density is a crucial number. Cosmologists denote it by the Greek letter Ω (omega). The fate of the universe depends on whether or not Ω exceeds one. At first sight our estimate of the actual average concentration of atoms in space seems to imply that Ω is only 1/25 (or 0.04), portending perpetual expansion, by a wide margin. But we should not jump too soon to that conclusion. We've come to realize in the last twenty years that there's a lot more in the universe than we actually see, such unseen material consisting mainly of 'dark stuff' of unknown nature. The things that shine – galaxies, stars and glowing gas clouds – are a small and atypical fraction of what is actually there, rather as the most conspicuous things in our terrestrial sky are cloud patterns, which are actually insubstantial vapours floating in the much denser clear air. Most of the material in the universe, and the main contributor to Ω, emits no light, nor infrared heat, nor radio waves, nor any other kind of radiation, and is consequently hard to detect.

The cumulative evidence for dark matter is now almost uncontestable. The way stars and galaxies are moving suggest that something invisible must be exerting a gravitational pull on them. This is the same line of argument by which we infer the existence of a black hole when a star is seen to be orbiting around an invisible companion; it's also the reasoning used in the nineteenth century when the planet Neptune was inferred to exist because the orbit of Uranus was deviated by the pull of a more distant unseen object.

In our Solar System, there is a balance between the tendency of gravity to make the planets fall towards the Sun, and the centrifugal effect of the orbital motions. Likewise, on the far bigger scale of an entire galaxy, there is a balance between gravity, which tends to pull everything together into the centre, and the disruptive effects of motion, which, if gravity didn't act, would make its constituent stars disperse. Dark matter is inferred to exist because the observed motions are surprisingly fast – too fast to be balanced just by the gravity of the stars and gas that we see.

We know how fast our Sun is circling around the central 'hub' of our galaxy; and we can measure the speeds of stars and gas clouds in other galaxies. These speeds, especially those of 'outliers' orbiting beyond most of the stars, are puzzlingly high. If the outermost gas and stars were feeling just the gravitational pull of what we can see, they should be escaping, just as Neptune and Pluto would escape from the Sun's influence if they were moving as fast as the Earth does. These high observed speeds tell us that a heavy invisible halo surrounds big galaxies – just as, if Pluto were moving as fast as the Earth (but were still in orbit rather than escaping), we would have to infer a heavy invisible shell outside the Earth's orbit but inside Pluto's.

If there weren't a lot of dark stuff, galaxies would not be stable but would fly apart. The beautiful pictures of discs or spirals portray what is essentially just 'luminous sediment' held in the gravitational clutch of vast swarms of invisible objects of quite unknown nature. Galaxies are ten times bigger and heavier than we used to think. The same argument applies, on a larger scale, to entire clusters of galaxies, each millions of light-years across. To hold them together requires the gravitational pull of about ten times more material than we actually see.

There is, of course, one assumption underlying these inferences of 'dark matter', namely that we know the force of gravity exerted by the objects we see. The internal motions within galaxies and clusters are slow compared with the speed of light, and so there are no 'relativistic' complications; we therefore just use Newton's inverse-square law, which tells us that if you move twice as far away from any mass then the force gets four times weaker. Some sceptics remind us that this law has only really been tested within our Solar System; it is plainly a leap of faith to apply it on scales a hundred million times larger. Indeed, we've now got tantalizing clues (see Chapter 10) that, on the scale of the entire universe, gravity is perhaps overwhelmed by another force that causes repulsion rather than attraction.

We should keep our minds open (or at least ajar) to the possibility that our ideas on gravity need reappraisal. If the force exerted at large distances were stronger than we would infer by extrapolating the inverse-square law – if it weren't four times weaker at twice the range – then clearly the case for dark matter would need rethinking. But we shouldn't abandon our theory of gravity without a struggle. We might be tempted to do so if there were no conceivable candidates for dark matter. However, there seem to be many options; only if these can *all* be ruled out should we, in my opinion, be prepared to jettison Newton and Einstein.[2]

There are other tell-tale signs of abundant 'dark matter'. All gravitating material, whether luminous or 'dark', deflects light rays, and so clusters can be 'weighed' by detecting how strongly they deviate the paths of light rays passing through them. Indeed, the deflection of starlight by the Sun's gravity, observed by Eddington and others during the 1919 total eclipse, famously offered an early test of relativity that propelled Einstein to world-wide celebrity. The Hubble Space Telescope has taken spectacular pictures of some clusters of galaxies lying about a billion light-years away. The pictures reveal a lot of faint streaks and arcs: each is a remote galaxy, several times further away than the cluster itself, whose image is, as it were, viewed through a distorting lens. Just as a regular pattern on background wallpaper looks streaky and distorted when viewed through a curved sheet of glass, the cluster acts like a 'lens' that focuses light passing through it. The visible galaxies in the cluster, all added together, aren't heavy enough to produce so much distortion. To bend the light so much, and cause such conspicuous distortion in the images of background galaxies, the cluster must contain ten times more mass than we see. These huge natural lenses offer a bonus to astronomers interested in how galaxies evolve, because they bring into view very remote galaxies that would otherwise be too faint to be seen.

We shouldn't really have been surprised to discover that dark matter, amounting to about ten times what we see, is the dominant gravitational influence on the cosmos. There's nothing implausible about dark matter *per se*: why *should* everything in the universe be shining? The challenge is to narrow down the range of candidates.

WHAT CAN THE DARK MATTER BE?

The inferred dark matter emits no light – indeed no radiation of any kind that we can detect. Nor does it absorb or scatter light. This means that it cannot be made of dust. We know that there is *some* dust in our galaxy, because starlight is scattered and attenuated by intervening clouds that are pervaded by tiny grains, rather like those that produce the haze from tobacco smoke. But if the grains cumulatively weighed enough to make up all the dark matter, they would black out our view of any distant stars.

Small faint stars are obvious suspects for the dark matter. Stars below eight per cent of the Sun's mass are called 'brown dwarfs'. They wouldn't be squeezed hot enough to ignite the nuclear fuel that keeps ordinary stars shining. Brown dwarfs definitely exist: some have been found as a by-product of searches for planets in orbit around brighter stars; others, especially nearby, have been detected by their very faint emission of red light. How many brown dwarfs might we expect altogether? Theory offers little guidance. The proportions of big and small stars are determined by very complicated processes that aren't yet understood. Not even the most powerful computers can tell us what happens when an interstellar cloud condenses into a population of stars; the processes are currently intractable, for the same reasons that weather prediction is so very difficult.

Individual brown dwarfs can be revealed by gravitational lensing. If one of them were to pass in front of a bright star, then the brown dwarf's gravity would focus the light, causing the bright star to appear magnified. As a consequence, a star would brighten up and fade in a distinctive way if a brown dwarf passed in front of it. This requires very precise alignment, and such events would consequently be very rare, even if there were enough brown dwarfs to make up all the dark matter in our galaxy. However, astronomers have carried out ambitious searches for these 'microlensing' events (called 'micro' to distinguish the phenomenon from the lensing by entire clusters of galaxies, as already mentioned). Millions of stars are monitored repeatedly in order to pick out those whose brightness changes from night to night. Many stars vary for all kinds of intrinsic reasons: some pulsate, some undergo flares, and some are orbiting around binary companions. The searches have found many thousands of these (which are interesting to some astronomers, though a tiresome complication for the microlensing searches). Occasionally, stars have been found to display the distinctive rise and fall in brightness that would be expected if an unseen mass had crossed in front of them and focused their light. It still isn't clear whether there are enough of these events to implicate a new 'brown dwarf' population, or whether ordinary faint stars, passing in front of brighter ones, are common enough to account for the events recorded.

There are several other candidates for dark matter. Cold 'planets' moving through interstellar space, unattached to any star, could exist in vast numbers without being detected; so could comet-like lumps of frozen hydrogen; so could black holes.

THE CASE FOR EXOTIC PARTICLES

Brown dwarfs or comets (or even black holes, if they are the remnants of dead stars) are, however, suspected to be only a minor constituent of the dark matter. This is because there are strong reasons for suspecting that dark matter isn't made of ordinary atoms at all. This argument is based on deuterium (heavy hydrogen).

As mentioned in the last chapter, any deuterium that we observe must have been made in the Big Bang, not in stars. The actual amount in our universe was, until recently, uncertain. But astronomers have detected the spectral imprint of deuterium, distinguishing it from ordinary hydrogen, in the light received from very distant galaxies. This measurement has needed the light-collecting power of new telescopes with ten-metre-diameter mirrors. The observed abundance is just a trace – only one atom in 50,000 is a deuterium atom. The proportion that should emerge from the Big Bang depends on how dense the universe is, and observations agree with theory if there are 0.2 hydrogen atoms in each cubic metre. This accords quite well with the actual number of atoms in objects that shine – half are in galaxies, and the other half is in intergalactic gas – but nothing much is then left over for the dark matter.

If there were enough atoms to make up all the dark matter –

which would imply at least five (and perhaps ten) times more than we actually see – the concordance with theory would be shattered. The Big Bang calculations would then predict *even less* deuterium, and *somewhat more* helium, than we actually observe: the origin of the deuterium in the universe would then become a mystery. This tells us something very important: the atoms in the universe, with a density of 0.2 per cubic metre, contribute only four per cent of the critical density, and the dominant dark matter is made of something that is inert as far as nuclear reactions are concerned. Exotic particles – not anything made of ordinary atoms at all – make the main contribution to Ω.[3]

The elusive particles called neutrinos are one option. They have no electric charge, and hardly interact at all with ordinary atoms: almost all neutrinos that hit the Earth go straight through. During the very first second after the Big Bang, when the temperature exceeded ten billion degrees, everything was so compressed that the reactions converting photons (quanta of radiation) into neutrinos would have been fast enough to come into balance. In consequence, the number of neutrinos left over from the 'cosmic fireball' should be linked to the number of photons. One can calculate, using physics that is quite standard and uncontroversial, that there should be 3/11 as many neutrinos as there are photons. There are now 412 million photons per cubic metre in the radiation left over from the Big Bang. There are three different species of neutrinos, and there would be 113 of each species in every cubic centimetre – in other words, hundreds of millions of neutrinos for every atom in the universe. It is of course the heaviest of the three species that is important in the dark matter context.

Because neutrinos so greatly outnumber atoms, they could be the dominant dark matter even if each weighed only a hundred-millionth as much as an atom. Before the 1980s, almost everyone believed neutrinos were 'zero rest-mass' particles; they would then carry energy and move at the speed of light, but their gravitational effects would be unimportant. (Likewise, the photons left over from the early universe, now detected as the microwave background radiation, don't now exert any significant gravitational effects.) But it now seems that neutrinos may weigh something, even though it is a very tiny amount indeed.

The best evidence for neutrino masses comes from the Kamiokande experiment in Japan, using a huge tank in a former zinc mine. The experimenters studied neutrinos that come from the Sun (where they are a by-product of the nuclear reactions in the central core), as well as others that are produced by very fast particles ('cosmic rays') impacting on the Earth's upper atmosphere. The experiments imply a non-zero mass, but one that is probably too small to render them important for the dark matter.[4] This is, nonetheless, a pivotal discovery about neutrinos themselves. At first sight it makes the microworld seem more complicated, but the masses may offer extra clues to the relation between neutrinos and other particles.

At least we know that neutrinos exist, although we don't yet know their exact masses. But there is a long list of hypothetical particles that *might* exist, and (if so) could have survived from the Big Bang in sufficient numbers to provide the dominant contribution to Ω. There are no very convincing arguments about how heavy each particle might be: best guesses suggest a hundred times as much as a hydrogen atom. If there were enough such particles to make up all the dark matter in our galaxy, there would be several thousand per cubic metre in the neighbourhood of the Sun; they would be moving at about the same speed as the average star in our galaxy – maybe 300 kilometres per second.

These particles, heavy but electrically neutral, would generally, like neutrinos, go straight through the Earth. However, a tiny proportion are likely to interact with an atom in the material they pass through. There would be only a few collisions per day within each of us (even though our bodies each contain nearly 10^{29} atoms). We ourselves clearly feel nothing. However, very sensitive experiments can detect the minuscule 'kick', or recoil, when such an impact happens in a lump of silicon or similar material. The detectors must be cooled to a very low temperature and placed deep underground (for instance, they are set up in a mine in Yorkshire, and in a tunnel under an Italian mountain) so as to reduce the confusion from other kinds of event that could drown out any genuine signal from dark-matter impacts.

Several groups of physicists have taken up the challenge of this 'underground astronomy'. It's delicate and tedious work, but if they succeed, they will not only find out what our universe is mainly made of but as a bonus they may discover an important new kind of particle. Only an extreme optimist would bet more than evens on success. This is because, at the moment, we have no theory that tells us what the dark-matter particles are and it's therefore hard to focus the search optimally. The next step in our theoretical understanding of sub-nuclear physics may involve a concept called 'supersymmetry', which aims to relate the nuclear force to the other forces within atoms (and thereby give us a better understanding of our cosmic number ε). Integral to this concept are some new kinds of electrically-neutral particles that would have been made in the Big Bang, and whose masses might be calculable.

Many other ideas are currently being considered. Some theorists favour a type of even lighter particle called an axion. Others suspect that the particles could be a billion times heavier than those currently being searched for (in which case there would be a billion times *fewer*, making detection even harder). Or they could be more exotic still – for instance, atom-sized black holes made in the ultra-high pressures of the early universe.

NARROWING DOWN THE OPTIONS

Some options for the dark matter can be ruled out; serious searches for other candidates, by a variety of techniques, are under way. Gravitational microlensing may detect enough faint stars or black holes. Experimenters at the bottom of mineshafts may detect some new kind of particle that pervades our galactic halo. Even negative results can sometimes be interesting because they exclude some tenable options.

There may well be several different kinds of dark matter. It would, for instance, be surprising if there weren't *some* brown dwarfs and black holes. However, exotic particles seem far more likely, because of the evidence from deuterium that most dark matter isn't made up of ordinary atoms.

It's embarrassing that more than ninety per cent of the universe remains unaccounted for – even worse when we realize that the dark matter could be made up of entities with masses ranged from 10^{-33} grams (neutrinos) up to 10^{39} gm (heavy black holes), an uncertainty of more than seventy powers of ten. This key issue may yield to a three-pronged attack:

1. The entities making up the dark matter may be directly detectable. Brown dwarfs may cause gravitational lensing of stars. If the dark matter in our galaxy is a swarm of particles, some of these might be detected by intrepid experimenters deep underground. I'm optimistic that if I were writing in five years' time, I would be able to report what the dark matter is.

2. Experimenters and theorists are already telling us more about neutrinos. It's possible (though it now seems unlikely) that neutrinos have enough mass to be an important dark-matter constituent. When the physics of extreme energies and densities is better understood, we should know what other kinds of particles might once have existed, and be able to calculate how these particles would have survived from the first millisecond of the universe just as confidently as we can now predict the amount of helium and deuterium surviving from the first three minutes.

3. Dark matter dominates galaxies. When and how galaxies formed, and the way that they are clustered, plainly

depends on what their gravitationally-dominant constituent is and how it behaves as the universe expands. We can make different guesses about the dark matter, calculate the outcome of each, and see which outcome most resembles what we actually observe. Such calculations (described in Chapter 8) can offer indirect clues to what the dark matter is.

WHY MATTER AND NOT ANTIMATTER?

We don't know yet what types of particle might have existed in the ultra-early phases of the universe nor how many survive. If, as I believe, the main contribution to Ω comes from new kinds of particle, our cosmic modesty may have to go a stage further. We are used to the post-Copernican idea that we don't occupy a special central place in the cosmos, but we must now abandon 'particle chauvinism' as well. The atoms that comprise our bodies and that make all visible stars and galaxies, are mere trace-constituents of a universe whose large-scale structure is controlled by some quite different (and invisible) substance. We see, as it were, just the white foam on the wave-crests, not the massive waves themselves. We must envisage our cosmic habitat as a dark place, made mainly of quite unknown material.

Ordinary atoms seem to be a 'minority' constituent of the universe, swamped by quite different kinds of particles surviving from the initial instants of the Big Bang. But it is actually more of a puzzle to understand why there are *any* atoms – why our universe isn't *solely* composed of dark matter.

To every kind of particle there is a corresponding antiparticle. There are protons (made up of three so-called 'quarks') and antiprotons (made up of three antiquarks); the 'anti' of an electron is a positron. Antiparticles annihilate when they encounter ordinary particles, converting their energy (mc^2) into radiation. No antimatter exists in bulk anywhere in or on the Earth. Tiny amounts can be made in accelerators, where particles are crashed together with sufficient energy to make extra particle-antiparticle pairs. Antimatter would be the ideal rocket fuel. When it annihilates, its *entire* rest-mass energy is released, compared with the fraction $\varepsilon = 0.007$ for rockets powered by nuclear fusion. Antimatter can survive only if 'quarantined' from ordinary matter; otherwise it betrays itself by generating intense gamma rays when it annihilates. We can be sure that our entire galaxy – all its constituent stars and gas – is matter rather than antimatter: its content is constantly being churned up and recycled by stellar births and deaths, and had it started off half matter and half antimatter there would by now be nothing left. But on much larger scales the mixing would be less efficient: we can't, for instance, refute the conjecture that 'superclusters' of galaxies consist alternately of matter and antimatter. So why is there a seeming bias in favour of one kind of matter?

There are 10^{78} atoms within our observable universe (mainly hydrogen atoms, each composed of a proton and an electron), but there do not seem to be so many antiatoms. The simplest universe, one might imagine, would have started off with particles and antiparticles mixed up in equal numbers. Our universe luckily wasn't like that. If it had been, then all protons would have annihilated with antiprotons during the dense early stages; it would have ended up full of radiation and dark matter but containing no atoms, no stars and no galaxies.

Why this asymmetry? The full 10^{78} excess could have been there right from the beginning, but this seems an unnaturally large number to accept as simply a part of the 'initial conditions'. The Russian physicist Andrei Sakharov is most widely famed for his role in developing the H-bomb, and later as a leading dissident in the final years of the Soviet Union; but he also contributed prescient ideas to cosmology. In 1967 he explored whether, during the cooling immediately after the Big Bang, a small asymmetry might favour particles over their antiparticles. This imbalance could create a slight excess of quarks over antiquarks (which would later translate into an excess of protons over antiprotons).

Sakharov's idea obviously requires some departure from perfect symmetry between the behaviour of matter and antimatter. Evidence for such an effect – a big surprise at the time – came in 1964 from two American physicists, James Cronin and Val Fitch, who were studying the decays of an unstable particle called a K^o. They found that this particle and its antiparticle weren't perfect mirror images of each other, but decayed at slightly different rates; some slight asymmetry was built in to the laws governing the decays. (This means, incidentally, that if we achieved contact with an 'alien' physicist who could report experiments done on another galaxy, we could tell whether that physicist was made of matter or antimatter – something that it would be prudent to check before planning a rendezvous!) The K^o decay involves only the so-called 'weak' force (which governs radioactivity and neutrinos) and not the strong nuclear force. In a unified theory of the forces, however, this type of asymmetry would 'carry over' from one force to the other, offering a basis for Sakharov's idea.

Suppose that, for every 10^9 quark-antiquark pairs, such an asymmetry had led to one extra quark. As the universe cooled, antiquarks would all annihilate with quarks, eventually giving quanta of radiation. This radiation, now cooled to very low energies, constitutes the 2.7 degree background heat pervading intergalactic space. But for every billion quarks that were annihilated with antiquarks, one would survive because it couldn't find a partner to annihilate with. There are indeed more than a billion times more radiation quanta (photons) in the universe than there are protons (412 million photons in each cubic metre, compared with about 0.2 protons). So all the atoms in the universe could result from a tiny bias in favour of matter over antimatter. We, and the visible universe around us, may exist only because of a *difference in the ninth decimal place* between the numbers of quarks and of antiquarks.

Our universe contains atoms and not antiatoms because of a slight 'favouritism' that prevailed at some very early stage. This implies, of course, that a proton (or its constituent quarks) can sometimes appear or disappear without the same thing happening to an antiproton. There is a contrast here with net electrical charge: this is *exactly* conserved, so that if our universe started off uncharged, there would always be an exact cancellation between positive and negative charges.

Atoms don't live for ever, although the decay rate appears to be incredibly low: a best guess for an atom's lifetime might be about 10^{35} years. This would mean that, on average, one atom would decay every year within a tank containing a thousand tons of water. Experiments in the same large underground tanks that are used to detect neutrinos cannot quite reach this sensitivity, but already tell us that the lifetime is at least 10^{33} years.

In the remote future, all stars will turn into cold white dwarfs, neutron stars or black holes. But the white dwarfs and neutron stars will themselves erode away as the constituent atoms decay. If this erosion took 10^{35} years, the heat generated by the prolonged decay would make each star radiate as much as a household electric heater. These feeble emitters would be the prime warmth (except for occasional flashes following stellar collisions) in the remote future, when all stars had exhausted their nuclear energy.

THE TUNING OF THE INITIAL EXPANSION

Ω may not be exactly one, but it is now at least 0.3. At first sight, this may not seem to indicate fine tuning. However, it implies that Ω was *very close indeed* to unity in early eras. This is because, unless expansion energy and gravitational energy are in exact balance (in which case Ω is, and remains, exactly equal to unity), the gap between those two energies widens: if Ω were to start off slightly less than unity in the early universe, eventually the kinetic energy would completely dominate (so that Ω becomes very small indeed); on the other hand, if Ω were substantially more than unity, then gravity would soon get the upper hand and bring the expansion to a halt.

The range of 'trajectories' for our actual universe, consistent with what the dark matter evidence tells us about the present value of Ω, is shown in Figure 6.1. The figure also depicts some

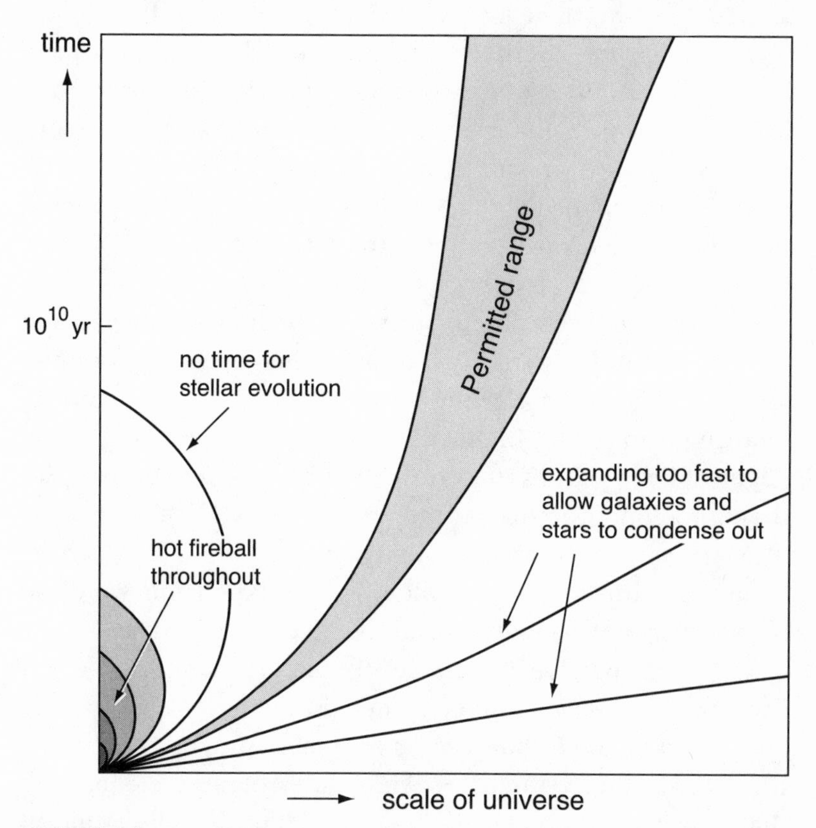

FIGURE 6.1
This diagram indicates various trajectories for possible universes. Despite the uncertainty in the present value of Ω, the initial conditions must have been tuned with remarkable precision in order for our universe to end up in the permitted range. Without this tuning, the expansion would either have been so fast that no galaxies could form, or so slow that the universe recollapsed before there was time for any interesting evolution. Explanations for this tuning are discussed in Chapter 9.

universes in which life as we know it *couldn't* have emerged. It highlights a basic mystery: Why is our universe still, after ten billion years, expanding with a value of Ω not too different from unity?

There are, as we've seen in the last chapter, good grounds for extrapolating back to when the universe was one second old and at a temperature of ten billion degrees. Suppose that you were 'setting up' a universe then. The trajectory it would follow would depend on the impetus it was given. If it were started *too fast*, then the expansion energy would, early on, have become so dominant (in other words, Ω would have become so small) that galaxies and stars would never have been able to pull themselves together via gravity and condense out; the universe would expand for ever, but there would be no chance of life. On the other hand, the expansion must not have been too slow: otherwise the universe would have recollapsed too quickly to a Big Crunch.

Any emergent complexity must feed on non-uniformities in density and in temperature (our own biosphere, for example, energizes itself by absorbing the Sun's 'hot' radiation and re-emitting it into cold interstellar space). Without being to the slightest degree anthropocentric in our concept of life, we can therefore conclude that a universe has to expand out of its 'fireball' state, and at least cool down below 3000 degrees, before any life can begin. If the initial expansion were too slow to permit this, there would be no chance for life.

In this perspective, it looks surprising that our universe was initiated with a very finely-tuned impetus, almost exactly enough to balance the decelerating tendency of gravity. It's like sitting at the bottom of a well and throwing a stone up so that it just comes to a halt exactly at the top – the required precision is astonishing: at one second after the Big Bang, Ω cannot have differed from unity by more than one part in a million billion (one in 10^{15}) in order that the universe should now, after ten billion years, be still expanding and with a value of Ω that has certainly not departed wildly from unity.

We have already noted that any complex cosmos must

incorporate a 'large number' $\mathcal{N}$ reflecting the weakness of gravity, and must also have a value of ε that allows nuclear and chemical processes to take place. But these conditions, though necessary, are not sufficient. Only a universe with a 'finely tuned' expansion rate can provide the arena for these processes to unfold. So Ω must be added to our list of crucial numbers. It had to be tuned amazingly close to unity in the early universe. If expansion was too fast, gravity could never pull regions together to make stars or galaxies; if the initial impetus were insufficient, a premature Big Crunch would quench evolution when it had barely begun.

Cosmologists react to this 'tuning' in different ways. The most common reaction seems, at first sight, perverse. This is to argue that because our early universe was set up with Ω very close to unity, there must be some deep reason why it is *exactly* one; in other words, because the 'tuning' is very precise, it must be absolutely perfect. This odd-looking style of reasoning has actually served well in other contexts; for instance, we know that in a hydrogen atom, the positive electric charge on the proton is cancelled by the negative charge on the orbiting electron, to immense precision – better than one part in 10^{21}. No measurement can, however, tell us that the net charge on an atom is *exactly* zero: there is always some margin of error. So-called 'grand unified theories', which interrelate electrical forces with nuclear forces, have, within the last twenty years, suggested a deep reason why the cancellation is exact. However, most physicists even fifty years ago would have guessed that the cancellation was exact, even though there weren't then any convincing arguments.

Another surprise is that the expansion rate (the Hubble constant) is the same in all directions: it can be described by a single 'scale factor', depicting the lengthening of the rods in Escher's lattice – see Figure 5.1. We could easily imagine a universe where the stretching was faster in some directions than in others. A less-uniform universe would seem to have more options open to it. Why, when we observe remote regions in opposite directions, do they look so similar and synchronized? Or why is the temperature of the background radiation, which has not been scattered since the temperature was 3000 degrees, almost the same all over the sky? As we shall see in Chapter 9, there is an attractive explanation – invoking a so-called 'inflationary phase' – for these features of our universe, and for the fine tuning of Ω in the early universe.

CHAPTER 7

THE NUMBER λ: IS COSMIC EXPANSION SLOWING OR SPEEDING?

The universe may
Be as large as they say.
But it wouldn't be missed
If it didn't exist.
Piet Hein

SEEING BACK INTO THE PAST

Our universe contains more mass in dark matter than in ordinary atoms. But is there enough to provide the full 'critical density' – to make Ω exactly equal to unity? The inferred amount within galaxies and clusters of galaxies falls short of this. However, dark matter *uniformly spread* through the universe would not influence the internal motions within clusters, nor the light-bending due to clusters, which magnifies and distorts the images of very distant galaxies. It would therefore be even more elusive. The extra material would only betray its presence by affecting the overall cosmic expansion. Can we, therefore, discover how the expansion rate is changing?

This is certainly possible in principle. The redshift of a distant object tells us how it was moving when its light set out, as opposed to how it is moving now. By observing the redshifts and distances of a remote population of galaxies (or any other type of object) we can therefore infer the expansion rate at an earlier era. Comparison with the present rate then tells us how much (if at all) the expansion rate has been changing.

Any change in the expansion rate would be so gradual that it would only show up over a 'baseline' of several billion years, so there is no hope of detecting it unless we can observe objects several billion light-years away. This isn't in itself an impediment, because superbly instrumented telescopes with ten-metre mirrors are now probing back to when the universe was no more than a tenth of its present age. More serious is the problem of finding distant objects that are sufficiently standardized, and allowing for the possibility that they look intrinsically different from their nearby counterparts because they are being observed at an earlier stage in their evolution.

The easiest objects to detect at high redshifts are 'quasars', the hyperactive centres of galaxies. They are very far from being 'standard candles': quasars with similar redshifts (in other words, at similar distances) display a wide range of apparent brightness. Even worse, they are so poorly understood that we do not know how their intrinsic properties might change as the universe gets older.

Galaxies themselves are somewhat better understood than quasars (though not as luminous), and we can now see them out to equally large redshifts, but here too there are problems. There is a whole zoo of different types, which are hard to classify. And they evolve as they age. They do this for several reasons: the existing stars evolve and die; new stars form from gas; or stars are added to the galaxy because it captures smaller neighbours (this is called 'galactic cannibalism').

Galaxies are too complicated, too varied and still too poorly understood to serve as 'standard candles'. They are far less well-understood than individual stars. Single stars are far too faint to be detected at cosmological distances: our telescopes detect a whole galaxy by picking up the total light from its billions of constituent stars. But some stars, in their death-throes, explode as supernovae, and for a few days blaze nearly as brightly as a whole galaxy containing many billions of ordinary stars.

HUNTING DISTANT SUPERNOVAE

A distinctive type of supernova, technically known as a 'Type 1a', signals a sudden nuclear explosion in the centre of a dying star, when its burnt-out core gets above a particular threshold of mass and becomes unstable. It is, in effect, a nuclear bomb with a standard yield. The physics is fairly well understood, and the details need not concern us. What is important is that Type 1a supernovae can be regarded as 'standard candles', bright enough to be detected at great distances. From how bright they appear, it should be possible to infer reliable distances, and thereby (by measuring the redshift as well) to relate the expansion speed and distance at a past epoch. Cosmologists hoped that such measurements would distinguish between a small slowdown-rate (expected if the dark matter has all been accounted for) or the larger rate expected if – as many theorists suspected – there was enough extra dark matter to make up the full 'critical density' so that the universe resembled the simplest theoretical model.

These supernovae, incidentally, display another trend that relates directly to their redshift: the remotest and most red-shifted ones appear to flare up and fade more slowly than closer ones of the same type. This is exactly what we would expect: a clock on a receding object *should* run slow. If it sends out periodic 'beeps', the later ones have further to travel, and so the intervals between their arrival are lengthened.[1]

The brightening and fading of a supernova is itself like a clock, so a slowdown in the 'light curves', proportional to the redshift, is just what we would expect if they are receding. It would have no natural explanation in a static universe. This is the best counter to any suspicion that the redshift is due to some kind of 'tired light' effect.

Astronomy is, in sociological terms, a 'big science': it requires large and expensive equipment. But the research programmes themselves generally don't require industrial-style teamwork of a kind that is obligatory in, for instance, the laboratories that use big accelerators to study subnuclear particles. Astronomers can still be individualists, pursuing solo projects by competing for a few nights' observing time on big telescopes (or, of course, by doing something innovative with a small telescope, like the astronomers who first discovered planets around other stars). But the enterprise of using supernovae for cosmology requires prolonged effort by many collaborators, using several telescopes. The first challenge is to 'catch' some photons – faint traces of light – from a stellar explosion that occurred billions of years ago. Distant supernovae are picked out by surveying the same patches of sky repeatedly, looking for occasional transient points of light in remote galaxies. The searches are done with moderately-sized telescopes because the biggest instruments are in such demand that not enough time can be allocated to any single programme, even one as important as this. Each supernova must then be observed repeatedly, so as to plot out its 'light curve' and measure the apparent brightness as accurately as possible. This preferably requires a ten-metre telescope on the ground, or the Hubble Space Telescope. Analysing all the data, and assessing its reliability, is itself an elaborate task.

There is a natural tendency to suspend judgement on any novel scientific claim, especially when it is unexpected, until it has been corroborated by independent evidence. There is sometimes a frustrating delay before this happens. It was therefore fortunate that two separate teams dedicated themselves to the 'supernova cosmology project'. The first serious entrant into the field was Saul Perlmutter, a physicist based at the Lawrence Berkeley Laboratory in California. Perhaps because he didn't then have much background in astronomy, he wasn't deterred by the difficulties and began his involvement around 1990. He gradually attracted and inspired a group of collaborators, from the UK as well as the US. A second group, also international, assembled later; this latter group contained several researchers who had introduced new techniques (which were then adopted also by Perlmutter's group) to classify the supernovae into subclasses that were even more standardized.

By 1998, each team had discovered about a dozen distant supernovae and mustered enough confidence to announce provisional results. There was less deceleration than would be expected if Ω were equal to one. This in itself wasn't surprising – there was no evidence for enough dark matter to raise Ω above around 0.3 – though it went against a strong theoretical prejudice that the cosmos would be 'simpler' if Ω were exactly unity. But what *was* a surprise was that there seemed no deceleration at all – indeed, the expansion seemed to be *speeding up*. The US-based magazine *Science* rated this as the number-one scientific discovery of 1998 in any field of research.

These observations are right at the limits of what is possible with existing telescopes. Remote supernovae are so faint that it's hard to measure them accurately. Furthermore, some astronomers worry that an intervening 'fog' of dust could attenuate the light, making the supernovae seem further away than they actually are. Also, the 'bomb' may not be quite standardized: for instance, its yield may depend on the amount of carbon etc in the precursor star, which would be systematically lower in objects that formed when the universe was younger (in other words, those that we observe with the highest redshifts). But cross-checks are being made, and every month more supernovae are added to the sample.

AN ACCELERATING UNIVERSE?

An *acceleration* in the cosmic expansion implies something remarkable and unexpected about space itself: there must be an extra force that causes a 'cosmic repulsion' even in a vacuum. This force would be indiscernible in the Solar System; nor would it have any effect within our galaxy; but it could overwhelm gravity in the still more rarified environment of intergalactic space. Despite the gravitational pull of the dark matter (which, acting alone, would cause a gradual deceleration), the expansion could then actually be *speeding up*. And we have to add another crucial number to our list to describe the strength of this 'antigravity'.

We normally think of the vacuum as 'nothing'. But if one were to remove from a region of interstellar space the few particles that it contains, and even shield it from the radiation passing through it, and cool it to the absolute zero of temperature, the emptiness that's left may still exert some residual force. Einstein himself conjectured this. As early as 1917, soon after he had developed his theory of general relativity, he began to think how that theory might apply to the universe. At that time, astronomers only really knew about our own galaxy, and the natural presumption was that the universe was static – neither expanding nor contracting. Einstein found that a universe that was set up in a static state would immediately start to contract because everything in it attracts everything else. A universe couldn't persist in a static state unless an extra force counteracted gravity. So he added to his theory a new number, which he called the 'cosmological constant', and denoted by the Greek letter λ (lambda). Einstein's equations then allowed a static universe where, for a suitable value of λ, a cosmic repulsion exactly balanced gravity. This universe was finite but unbounded: any light beam that you transmitted would eventually return and hit the back of your head.

This so-called 'Einstein universe' became no more than a curiosity after 1929. Astronomers had by then realized that our galaxy was just one of many, and that distant galaxies were receding from us: the universe wasn't static, but was expanding. Einstein thereafter lost interest in λ. Indeed, George Gamow's autobiography *My World Line* recalls a conversation in which Einstein, three years before his death, rated λ as his 'biggest blunder', because if he hadn't introduced it, his equations would have obligated the conclusion that our universe would be expanding (or contracting). He could then maybe have predicted the expansion before Edwin Hubble discovered it.

Einstein's reason for inventing λ has been obsolete for seventy years. But that doesn't discredit the concept itself. On the contrary, λ now seems less contrived and *ad hoc* than Einstein thought it was. Empty space, we now realize, is anything but simple. All kinds of particles are latent in it. Any particle, together with its antiparticle, can be created by a suitable concentration of energy. On an even tinier scale, empty space may be a seething tangle of strings, manifesting structures in extra dimensions. From our modern perspective the puzzle is: Why is λ so small? Why don't all the complicated processes that are going on, even in empty space, have a net effect that is much larger? Why isn't space as dense as an atomic nucleus or a neutron star (in which case it would close up on itself within ten or twenty kilometres)? Or even, perhaps, why isn't space as dense as the universe was at 10^{-35} seconds – an era whose significance for unified theories is discussed in later chapters? In fact, it is lower than that ultra-early density by a factor of 10^{120} – perhaps the worst failure of an order-of-magnitude guess in the whole of science. The value of λ may not be exactly zero, but it is certainly so weak that it can only compete with the very dilute gravity of intergalactic space.

Some theorists have suggested that space has a complicated microstructure of tiny black holes that adjusts itself to compensate for any other energy in the vacuum, and leads to λ being exactly zero. If our universe is indeed accelerating, and λ is not zero, this would scupper such arguments and also caution us against the line of thought that 'because something is remarkably small, there must be some deep reason why it is exactly zero'.

THE CASE FOR A NON-ZERO λ

The case for a non-zero λ at the time of writing (Spring 1999) is strong but not overwhelming. There could be unsuspected trends or errors in the supernova observations that haven't been properly allowed for. But other evidence, albeit of a slightly technical and indirect kind, bolsters the case for an accelerating universe. The background radiation – the 'afterglow' surviving from the Big Bang – is not completely uniform across the sky; there is a slight patchiness in the temperature, caused by the non-uniformities that evolve into galaxies and clusters. The expected size of the most prominent patches can be calculated. How large they appear in the sky – whether, for instance, they are one degree across or two degrees across – depends on the amount of focusing by the gravity of everything along the line of sight. Measurements of this kind weren't achieved until the late 1990s (they are made from high dry mountain sites, from Antarctica, or from long-duration balloon flights) and they tell against a straightforward low-density universe. If Ω were really 0.3, and λ were exactly zero, the seeds of clusters would appear smaller than they actually do. However, any energy latent in the vacuum contributes to the focusing. If λ were around 0.7, we get a pleasant consistency with these results, as well as with the supernova evidence for accelerating expansion.

Gravity is the dominant force in planets, stars and galaxies. But on the still-larger scale of the universe itself, the average density is so low that a different force may take over. The cosmic number λ – describing the weakest force in nature, as well as the most mysterious – seems to control the universe's expansion and its eventual fate. Einstein's 'blunder' may prove a triumphant insight after all. If it does, it will not be the only instance in which his work has had an impact that he himself failed to foresee. The most remarkable implication of general relativity is that it predicted black holes; but his attitude was summarized thus by Freeman Dyson:[2]

> Einstein was not only sceptical, he was actively hostile, to the idea of black holes. He thought the black hole solution was a blemish to be removed from the theory by a better mathematical formulation, not a consequence to be tested by observation. He never expressed the slightest enthusiasm for black holes, either as a concept or a physical possibility.

If λ isn't zero, we are confronted with the problem of why it has the value we observe – one smaller, by very many powers of ten, than what seems its 'natural' value. Our present cosmic environment would be very little different if it were even smaller (though the long-range forecast, discussed below, would be somewhat altered). However, a much higher value of λ would have had catastrophic consequences: instead of becoming competitive with gravity only after galaxies have formed, a higher-valued λ would have overwhelmed gravity earlier on, during the higher-density stages. If λ started to dominate before galaxies had condensed out from the expanding universe, or if it provided a repulsion strong enough to disrupt them, then there would be no galaxies. Our existence requires that λ should not have been too large.

THE LONG-RANGE FUTURE

Geologists infer the Earth's history from strata in the rocks; climatologists can infer changes in temperature over the last million years by drilling through successive layers of Antarctic ice. Likewise, astronomers can study cosmic history by taking 'snapshots' of the galaxies at different distances: those more remote from us (with larger redshifts) are being viewed at earlier stages in their evolution. The challenge for theorists (see Chapter 8) is to understand galaxies and how they evolve, and to produce computer simulations that faithfully match the reality.

Most galaxies have now settled down into a sedate

maturity, an equilibrium where their 'metabolism' has slowed. Fewer new stars are forming, and few bright blue stars are shining. But what about the long-range future? What would happen if we came back when the universe was ten times older – a hundred billion rather than ten billion years old? My favoured guess (before there was much relevant evidence) used to be that the expansion would by then have halted and been succeeded by recollapse to a Big Crunch in which everything experienced the same fate as an astronaut who falls inside a black hole. Our universe would then have a finite timespan for its continued existence, as well as being bounded in space. But this scenario requires Ω to exceed unity in value, contrary to the evidence that has mounted up in recent years. Dark matter assuredly exists, but there does not seem to be enough to yield the full 'critical density': Ω seems to be less than unity. Furthermore, an extra cosmic repulsion, described by λ, may actually be speeding-up the expansion of our universe.

It seems likely that expansion will continue indefinitely. We can't predict what role life will have carved out for itself ten billion (or more) years hence: it could be extinct; on the other hand, it could have evolved to a state where it can influence the entire cosmos, perhaps even invalidating this forecast. But we can compute the eventual fate of the inanimate universe: even the slowest-burning stars would die, and all the galaxies in our Local Group – our Milky Way, Andromeda, and dozens of smaller galaxies – would merge into a single system. Most of the original gas would by then be tied up in the dead remnants of stars; some would be black holes; others would be very cold neutron stars or white dwarfs.

Looking still further ahead, processes far too slow to be discernible today could come into their own. Collisions between stars within a typical galaxy are immensely infrequent (fortunately for our Sun), but their number would mount up. The drawn-out terminal phases of our galaxy would be sporadically lit up by intense flares, each signalling a collision between two dead stars. The loss of energy via gravitational radiation (an effect predicted by Einstein's theory of general relativity) – imperceptibly slow today, except in a few binary stars where the orbits are specially close and fast – would, given enough time, grind down all stellar and planetary orbits. Even atoms may not live for ever. In consequence, white dwarfs and neutron stars will erode away because their constituent particles decay. Eventually, black holes will also decay. The surface of a hole is made slightly fuzzy by quantum effects, and it consequently radiates. In our present universe, this effect is too slow to be interesting unless mini-holes the size of atoms actually exist. The timescale is 10^{66} years for the total decay of a stellar-mass hole; and a hole weighing as much as a billion suns would erode away in 10^{93} years.

Eventually, after 10^{100} years have passed, the only surviving vestige of our Local Group of galaxies would be just a swarm of dark matter and a few electrons and positrons. All galaxies beyond our Local Group would undergo the same internal decay, and would move further from us. But the speed with which they disperse depends crucially on the value of λ. If λ were zero, the pull of ordinary gravity would slow down the recession: although galaxies would move inexorably further away, their speed (and redshift) would gradually diminish but never quite drop to zero. If our remote descendants had powerful-enough telescopes to detect highly redshifted galaxies, despite their intrinsic fading and ever-increasing remoteness, they would actually be able to detect more than are visible in our present sky. After (say) 100 billion years, we would be able to see out as far as 100 billion light-years; objects that are now far beyond our present horizon, because their light hasn't yet had time to reach us, would come into view.

But if λ isn't zero, the cosmic repulsion will push galaxies away from each other at an *accelerating* rate. They will fade from view even faster because their redshifts increase rather than diminish. Our range of vision will be bounded by a horizon that is rather like an inside-out version of the horizon around a black hole. When things fall into a black hole, they accelerate, getting more and more redshifted and fading from view as they approach the hole's 'surface'. A galaxy in a λ-dominated universe would accelerate away from us, moving ever closer to the speed of light as it approaches the horizon. At late times, we will not see any further than we do now. All galaxies (except Andromeda and the other small galaxies gravitationally bound into our own Local Group) would be fated to disappear from view. Their distant future lies beyond our horizon, as inaccessible to us as the events inside a black hole. Extragalactic space will become exponentially emptier as the aeons advance.

CHAPTER 8

PRIMORDIAL 'RIPPLES': THE NUMBER Q

The universe was brought into being in a less than fully formed state, but was gifted with the capacity to transform itself from unformed matter into a truly marvellous array of structure and life forms.

St Augustine

GRAVITY AND ENTROPY

In nature, as in music or painting, the most appealing patterns are neither completely regular and repetitive nor completely random and unpredictable, but they combine both these features. The elaborately structured cosmic environment that we see around us is not completely ordered; nor has it run down to an utterly random state. There are ninety-two different kinds of atoms in nature, rather than just the simple hydrogen, deuterium and helium that were forged in the Big Bang. Some of these atoms now find themselves in complex organisms in our Earth's biosphere; some are in stars; others are dispersed in the voids of intergalactic space. And the temperature contrasts are also immense: the stars have blazing surfaces (and still hotter centres), but the dark sky is close to the 'absolute zero' of temperature – warmed to just 2.7 degrees by the microwave afterglow from the Big Bang.

That this intricate complexity all emerged from a boringly

amorphous fireball might seem to violate a hallowed physical principle: the Second Law of Thermodynamics. This law describes an inexorable tendency towards uniformity, and away from patterns and structure: things tend to cool if they're hot, and to warm up if they're cold. Ink and water can readily mix, whereas the reverse process – stirring a murky liquid until the dye concentrates into a black drop – would astonish us. Ordered states get messed up, but not the reverse. In technical jargon, 'entropy' can never decrease. An apparent decrease locally is always outweighed by an entropy increase elsewhere. The classic example of this principle is a steam engine, where the ordered motion of a piston is always accompanied by wasted heat.

We need to rethink our intuitions, however, when gravity comes into play. Stars, for instance, are held together by the inward pull of their own gravity. This is balanced by the pressure of their hot interiors pushing out. Odd though it seems, stars *heat up* when they *lose* energy. Suppose that the fuel supply in the Sun's centre were switched off. Its surface would stay bright because heat diffuses from the even hotter core. If nuclear fusion didn't regenerate this heat, the Sun would gradually deflate as energy leaked away (within about ten million years, as Lord Kelvin realized in the nineteenth century). But this deflation would actually make the core *hotter* than before: gravity pulls more strongly at shorter distances, and the central temperature would have to rise in order to provide enough pressure to balance the greater force pressing down on it. Something similar happens when an artificial satellite gradually spirals in to a lower orbit because of atmospheric drag: it heats up, but only half the energy released from gravity goes into heat; the other half goes into *speeding up* the satellite (because a closer-in orbit is faster).

So it should not surprise us that new stars condense within irregular clouds of cool dusty gas. The densest regions contract because of their own gravity, becoming so compressed that they light up as stars. Exactly how this happens in, for instance, the Orion cloud or the Eagle Nebula, and the proportions of big and small stars that result from this process, are still too hard to calculate even with the biggest computers. (This is why we aren't sure how many brown dwarf stars there are, which could contribute to the dark matter in our galaxy.) But star formation poses no mystery in principle: once gravity gets a grip on a system, it inexorably contracts.

FROM THE BIG BANG TO GALAXIES

The gas clouds within our galaxy (and within others) have been churned and recycled so much that they retain no 'memory' of their origins. Star formation is therefore insensitive to the wider cosmos. But the emergence of the galaxies themselves is less straightforward than the equivalent process for stars. Their origin lies in the early universe; they are shaped by their 'genetics' as well as by their environment.

If our universe had started off completely smooth and uniform, it would have remained so throughout its expansion. After ten billion years, it would contain thinly spread dark matter, and hydrogen and helium gas so rarified that there was less than one atom in each cubic metre. It would be cold and dull: no galaxies, therefore no stars, no periodic table, no complexity, certainly no people. But even *very slight* irregularities in the early phases make a crucial difference, because density contrasts amplify during the expansion. Any patch slightly denser than average decelerates more, because it feels extra gravity; its expansion lags further and further behind that of an average region. (If, by analogy, we throw two balls upwards with slightly different speeds, their trajectories may, to start with, differ only imperceptibly. The slower ball, however, will have completely stopped, and already started to fall, while the faster is still moving upwards.) Gravity amplifies slight 'ripples' in an almost featureless fireball, enhancing the density contrasts until the overdense regions stop expanding and condense into structures held together by gravity.

The most conspicuous structures in the cosmos – stars, galaxies, and clusters of galaxies – are all held together by gravity. We can express how tightly they are bound together – or, equivalently, how much energy would be needed to break up and disperse them – as a proportion of their total 'rest-mass energy' (mc^2). For the biggest structures in our universe – clusters and superclusters – the answer is about one part in a hundred thousand. This is a pure number – a ratio of two energies – and we call it *Q*.

The fact that *Q* is so small (of the order of 10^{-5}) means that gravity is actually quite weak in galaxies and clusters. Newton's theory is therefore good enough for describing how the stars move within a galaxy, and how each galaxy traces out an orbit under the gravitational influence of all the other galaxies and the dark matter within a cluster. The smallness of *Q* also means that we can validly treat our universe as approximately homogeneous, just as we'd regard a globe as smooth and round if the height of the waves or ripples on its surface were only 1/100,000 of its radius (equivalent to only 60 metres for a globe the size of the Earth).

The ripples would have been imprinted very early on, before the universe 'knew' about galaxies and clusters; there would be nothing special about these sizes (or, indeed, about any dimensions that seemed significant in our present universe). The simplest guess would be that nothing in the early universe favours one scale rather than another, so that the ripples are the same on every scale. The degree of initial 'roughness' was somehow established when our entire universe was of microscopic size: how this could have happened is conjectured in the next chapter. The number *Q* is crucial for determining the 'texture' of structure in our universe, which would be very different if its value were either much larger or much smaller.

RIPPLES IN THE MICROWAVE AFTERGLOW

Our universe started off dense and opaque, like the glowing gas inside a star. But after half a million years of expansion, the temperature had dropped to around 3000 degrees – slightly cooler than the Sun's surface. As the universe cooled further, it literally entered a dark age. The darkness persisted until the first protogalaxies formed and lit it up again.

Probing how the dark age ended is a challenge for astronomers in the next decade. Much hope is placed in the proposed 'Next Generation Space Telescope'. This is planned to have sensitive detectors for red light and infrared radiation, and an eight-metre mirror (compared with only 2.4 metres for the Hubble Space Telescope).

The microwave background radiation, the afterglow from the Big Bang itself, is a direct message from an era when galaxies only existed 'in embryo'. Slightly overdense regions, expanding slower than average, were destined to become galaxies or clusters; others, slightly underdense, were destined to become voids. And the microwave temperature should bear the imprint of these fluctuations. The expected effect would be about one part in 100,000 – essentially the same number as Q, the fundamental number characterizing the ripple amplitude.

An undoubted cosmological triumph of the 1990s has been the actual mapping of these precursors of cosmic structure. The background microwave radiation is about a hundred times weaker than the emission from the Earth (whose surface temperature is about 300 degrees above absolute zero). The daunting technical challenge is to measure temperature differences *a hundred thousand times smaller still.* NASA's COBE satellite, launched in 1990, achieved outstanding accuracy in confirming that the microwaves had a 'black body' spectrum (see Chapter 5). It also carried the first instrument sensitive enough to discern that the radiation from some directions was slightly hotter than from others. It scanned the whole sky, measuring the temperature with enough precision to map its non-uniformities.

Measurements of this kind are best made from space because water vapour in the atmosphere absorbs some of the radiation. COBE has been followed up by further measurements, made from mountaintops, from the South Pole (where the water vapour is low) or from equipment flown in balloons. These new experiments can only map a small area – not the entire sky, as a satellite can – but they achieve the same sensitivity at enormously less expense.

The next big advance will, however, come from two spacecraft that will carry more advanced and sensitive sensors than COBE did: NASA's Microwave Anisotropy Probe (MAP) and the European Space Agency's Planck/Surveyor. These will, within a few years, yield precise-enough data on the 'roughness' of the early universe on many different scales, to settle key questions about how galaxies emerged. The microwave background carries a lot of information about the ultra-early universe. It will, for instance, help to pin down Ω and λ, as well as Q.

It was actually a relief rather than a surprise to find non-uniformities in the afterglow temperature at a level of one part in 100,000. If the background microwaves had implied an even smoother early universe, the clusters and superclusters in our present universe would have been a puzzle: there would need to have been some extra force, apart from gravity, that could enhance the density contrasts even faster.

But the fact that Q is *only* 1/100,000 is really the most remarkable feature of our universe. If you picked up a stone that was spherical to a precision of one part in 100,000, you might wonder what caused the small irregularities but you'd be even more perplexed by the overall smoothness. 'Inflation', described in Chapter 9, is the best theory we have of this, and the temperature fluctuations offer important tests of these ideas.

THE EVOLUTION OF 'VIRTUAL' UNIVERSES

When the universe was a million years old, everything was still expanding almost uniformly. How did the structures condense out, and develop into the cosmic scene we now observe? Nowadays we can use a computer to study 'virtual' universes. At the start of the simulation the material is expanding, but not quite uniformly because irregularities corresponding to the specified value of Q are fed in as part of the initial conditions.

The dominant gravitating stuff is the 'dark matter', particles surviving from the early universe that hardly ever collide with each other, but are influenced by gravity. If you averaged over larger and larger volumes, the early universe would have appeared increasingly smooth.[1] This means that, were gravity the only relevant force, small scales would condense first. Cosmic structure forms hierarchically, from the bottom up. Swarms of dark matter on subgalactic scales condense out first; these merge into galactic-mass objects, which then form clusters. It takes longer for gravity to reverse the expansion on larger scales.

But this hierarchical clustering in itself leads to a dark and sterile universe. The 'leaven' for the universe is the atoms. Their total mass is much less than that of the dark matter: they ride along passively, constituting a dilute gas that 'feels' the dark matter's gravity. But everything we actually see depends on this gas.

The gas behaves in a more complicated way than the dark matter, because gravity isn't the only force acting on it. Gas 'feels' gravity, but it exerts a pressure as well. This pressure prevents the gas from being pulled by gravity into very small 'clumps' of dark matter, but gravity wins on scales above a million solar masses. The first gaseous condensations to form – those that would cause the 'first light' that ends the cosmic dark age – are consequently a million times heavier than stars. The computer programs used to follow the gas motions resemble those used by aeronautical engineers to study flows around wings and through turbines. Such calculations are deemed reliable enough to be a substitute for wind-tunnel tests; but, even so, computing what happens inside one of these collapsing clouds is much harder, and nobody has yet performed a simulation that starts with a single cloud and ends up with a population of stars. A cloud containing a million solar masses of gas could fragment into a million separate stars like the Sun, or into fewer objects of larger mass. It could even remain in one piece, and contract into a single superstar or quasar.

These first objects would have formed when the universe was only a few hundred million years old – a few per cent of its present age. By the time the universe was a billion years old, galaxy-sized structures would have built up, each an assemblage of stars and held together not only by its own gravity but by the dark matter, which is configured in a 'swarm' ten times larger and heavier. Gas continues to fall inwards into these objects and to cool down. If it is spinning, the gas settles into a disc, and condenses into stars, thereby initiating the recycling process that synthesizes and disperses all the elements of the periodic table.

Computer simulations that show at least the broad outline of these processes can be run as movies, depicting the expansion of our universe and the emergence of galaxies about sixteen powers of ten faster than actually happened! Figure 8.1 shows six frames from one such simulation.

Like the individual galaxies, clusters and superclusters are the outcome of gravitational aggregation. The newly formed galaxies would not have been spread completely uniformly – there would be slightly more in some places than in others. As the expansion continued, regions containing excess mass would suffer extra deceleration, so that the galaxies in those regions ended up conspicuously more closely packed than average.[2]

How can we check whether a virtual universe is indeed an accurate resemblance to our real one? The simulation must

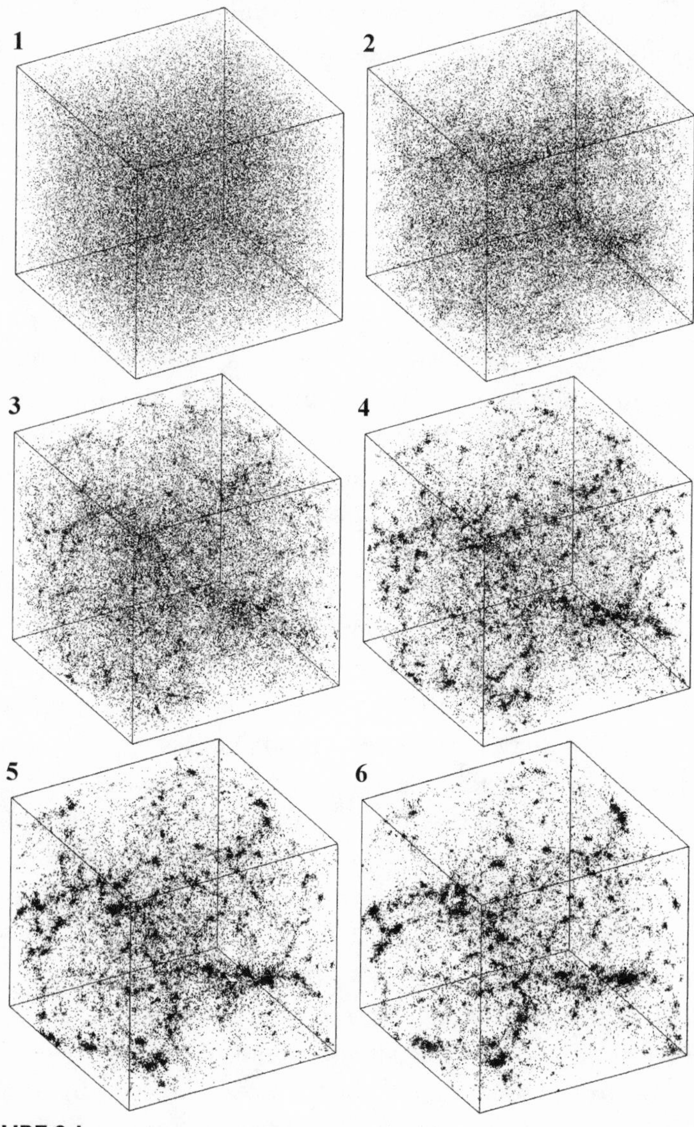

FIGURE 8.1

Six frames from a computer simulation showing how structure emerges in the expanding universe. In these pictures, the overall expansion is subtracted out, so that the boxes remain the same size. Initially, the incipient structure consists of barely perceptible irregularities. During the expansion, overdense regions lag further and further behind. Density contrasts grow, eventually condensing out to form gravitationally bound structures. These structures merge together, producing the galaxies – a prerequisite for our own emergence.

mimic the observed properties of galaxies today – their characteristic sizes and shapes, the proportions that are disc-like and the proportions that are elliptical – and the way that they are clustered. But it must do more: it must match the 'snapshots' that tell us what galaxies were like, and how they were clustered, at earlier times.

As discussed earlier, the light now reaching us from the remotest galaxies (and which new-generation telescopes can detect and analyse) set out when they were newly formed. And they look different from present-day galaxies. None has yet settled down into steadily spinning discs, and only a small fraction of their constituent gas has yet turned into stars. Most are small: it took successive mergers, and cannibalism by dominant galaxies of their smaller neighbours, to build up the large ones that we see today.

As a by-product of early star formation, something even more interesting happens. Some of the gas settles into the centre of the swarm of dark-matter particles, contracts under its own gravity, and builds up into a 'superstar' more than a million times heavier than an ordinary star. Such a big object shines so brightly that its nuclear fuel doesn't last long; it ends its life not by exploding but by collapsing to form a black hole. Thus, once galaxy formation starts, space gets 'punctured' by these holes. Gas continues to fall into them, releasing a power that outshines the rest of the galaxy.

These objects are called 'quasars', or 'active galactic nuclei', and they are interesting for two reasons. First, they shine more brightly than the galaxies themselves, and therefore serve as probes to illuminate the remote universe. Spectra of quasar light reveal clouds of gas along the line of sight, and yield our best evidence to date for the amount of deuterium – an important check, as we have seen, on the Big Bang theory. Secondly they permit important tests of Einstein's theory of general relativity. The power they emit comes from material that is swirling very close to a black hole, and perhaps even from the spinning hole itself. There is no real chance of getting an actual image of this flow – it would be even more of a challenge than imaging an Earth-like planet around another star – but the radiation it emits is redshifted by the strong gravity (and this would be additional, of course, to the ordinary cosmological redshift). There would also be large Doppler shifts because of the high speed with which the gas swirls around near the hole (red on the side that is moving away; blue from the approaching gas on the other side). From the inferred motions and gravitational fields, we can test whether black holes have the actual exact properties that Einstein's theory predicts.

HOW MUCH IS PREDICTABLE?

If one had to summarize, in just one sentence, 'What's been happening since the Big Bang?', the best answer might be to take a deep breath and say: 'Ever since the beginning, gravity has been moulding cosmic structures and enhancing temperature contrasts, a prerequisite for the emergence of the complexity that lies around us ten billion years later, and of which we are part.'

Once systems form that are heavy enough to be self-gravitating, departures from equilibrium grow. Our universe can thus have evolved from a primordial fireball, uniformly hot, into a structured state containing very hot stars radiating into very cold empty space. This sets the stage for increasingly intricate cosmic evolution, and the emergence of life. Individual stars become denser as they evolve (some ending as neutron stars or black holes), whereas overall the matter gets more thinly spread. These complexities are the outcome of a chain of events that cosmologists can trace back to an ultra-dense primal medium that was almost structureless.

Our view of how cosmic structure emerged is, like the Darwinian view of biological evolution, a compelling general scheme. As with Darwinism, how the whole process got started is still a mystery: the way Q is determined (perhaps as microscopic vibrations in the ultra-early universe) is still perplexing, just as the origin of the first organisms on Earth is. But cosmology is simpler in one important respect: once the starting point is specified, the outcome is in broad terms predictable. All large patches of the universe that start off the same way end up statistically similar. In contrast, the gross course of biological evolution is sensitive to 'accidents' – climatic changes, asteroid impacts, epidemics and so forth – so that, if the Earth's history were rerun, it could end up with a quite different biosphere.

That's why computer simulations of structure formation are so important. Galaxies and clusters are the outcome of gravity acting on initial irregularities. We don't try to explain the detailed pattern, only the statistics – just as an oceanographer aims to understand the statistics of waves, not the details of a wave in a single snapshot at a particular place and time.

The starting point is an expanding universe, described by Ω, λ and *Q*. The outcome depends sensitively on these three key numbers, imprinted (we are not sure how) in the very early universe.

THE TUNING OF *Q*

The formation of galaxies, clusters and superclusters obviously requires the universe to contain enough dark matter and enough atoms. The value of Ω must not be too low: in a universe that contained radiation and very little else, gravity could never overwhelm pressure. And λ mustn't be so high that the cosmic repulsion overwhelms gravity before galaxies have formed. There must also be enough ordinary atoms, initially in diffuse gas, to form all of the stars in all of the galaxies. But we've seen that something else is needed as well, namely initial irregularities to 'seed' the growth of structure. The number *Q* measures the amplitude of these irregularities or 'ripples'. Why *Q* is about 10^{-5} is still a mystery. But its value is crucial: were it much smaller, or much bigger, the 'texture' of the universe would be quite different, and less conducive to the emergence of life forms.

If *Q* were *smaller* than 10^{-5} but the other cosmic numbers were unchanged, aggregations in the dark matter would take longer to develop and would be smaller and looser. The resultant galaxies would be anaemic structures, in which star formation would be slow and inefficient, and 'processed' material would be blown out of the galaxy rather than being recycled into new stars that could form planetary systems. If *Q* were smaller than 10^{-6}, gas would never condense into gravitationally bound structures at all, and such a universe would remain forever dark and featureless, even if its initial 'mix' of atoms, dark matter and radiation were the same as in our own.

On the other hand, a universe where *Q* were substantially *larger* than 10^{-5} – where the initial 'ripples' were replaced by large-amplitude waves – would be a turbulent and violent place. Regions far bigger than galaxies would condense early in its history. They wouldn't fragment into stars but would instead collapse into vast black holes, each much heavier than an entire cluster of galaxies in our universe. Any surviving gas would get so hot that it would emit intense X-rays and gamma rays. Galaxies (even if they managed to form) would be much more tightly bound than the actual galaxies in our universe. Stars would be packed too close together and buffeted too frequently to retain stable planetary systems. (For similar reasons, solar systems are not able to exist very close to the centre of our own galaxy, where the stars are in a close-packed swarm compared with our less-central locality).

The fact that *Q* is 1/100,000 incidentally also makes our universe much easier for cosmologists to understand than would be the case if *Q* were larger. A small *Q* guarantees that the structures are all small compared with the horizon, and so our field of view is large enough to encompass many independent patches each big enough to be a fair sample. If

Q were much bigger, superclusters would themselves be clustered into structures that stretched up to the scale of the horizon (rather than, as in our universe, being restricted to about one per cent of that scale). It would then make no sense to talk about the average 'smoothed-out' properties of our observable universe, and we wouldn't even be able to define numbers such as Ω.

The smallness of Q, without which cosmologists would have made no progress, seemed until recently a gratifying contingency. Only now are we coming to realize that this isn't just a convenience for cosmologists, but that life couldn't have evolved if our universe didn't have this simplifying feature.

CHAPTER 9

OUR COSMIC HABITAT III: WHAT LIES BEYOND OUR HORIZON?

> Then assuredly the world was made, not in time, but simultaneous with time. For that which is made in time is made both after and before some time – after that which is past, before that which is future. But none could then be past, for there was no creature by whose movements its duration could be measured. But simultaneously with time the world was made.
>
> St Augustine

HOW BELIEVABLE IS THE BIG BANG STORY?

The Big Bang theory has lived dangerously for more than thirty years. Various measurements could have refuted it if they had turned out differently. Here are five of them:

- Astronomers might have discovered an object whose helium abundance was zero, or at any rate well below 23 per cent of that of hydrogen. This would have been fatal, because fusion of hydrogen in stars can readily boost helium *above* its pre-galactic abundance but there is no way of converting all the helium back to hydrogen.

- The background radiation measured so accurately by COBE might have turned out to have a spectrum that differed from the expected 'black body' or thermal form.[1]

- Physicists might have discovered something about *neutrinos* that was incompatible with the Big Bang. In the 'fireball', neutrinos would outnumber the atoms by a huge factor – around a billion – just as the photons do. If each neutrino weighed even a millionth as much as an atom, they would, in total, contribute too much mass to the present universe – more, even, than could be hidden in dark matter. As discussed in Chapter 6, the actual masses (if not zero) seem to be too low to embarrass the theory. But they could have turned out higher.
- The deuterium abundance could have been out of line with the amount expected to survive from the Big Bang.
- The temperature fluctuations over the sky could have implied a value of *Q* that was incompatible with what is inferred from the present-day structure in the universe, rather than, as discussed in Chapter 8, being consistent with a value of 1/100,000.

The Big Bang theory has survived these tests. The grounds for extrapolating back to the stage when our universe had been expanding for a second (when the helium began to form) deserve to be taken as seriously as, for instance, inferences from rocks and fossils about the early history of our Earth, which are equally indirect (and less quantitative).

Perhaps we can deepen our understanding, and even 'explain' the key cosmic numbers, by extrapolating still further back – not just into the first second but into the first tiny fraction of a second.

We can confidently go back a bit closer to the Big Bang, but not much. For the first millisecond we are less sure of the physics because everything would have been denser than a neutron star. Very hot and dense conditions can be simulated, on a microscopic scale, by experiments that crash together very energetic particles. But there are limits to how far back this technique can take us. Not even the giant Large Hadron Collider, being built at CERN in Geneva, will achieve the

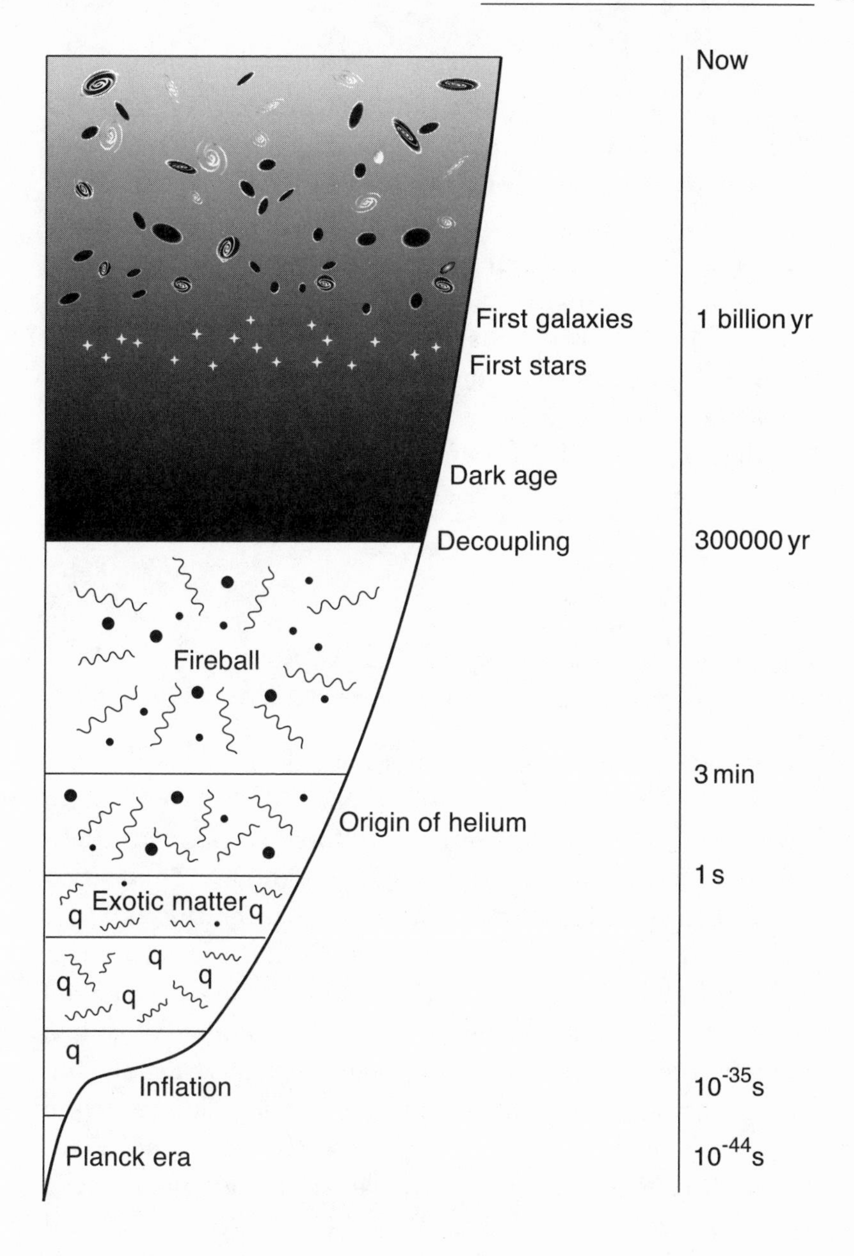

FIGURE 9.1
A time chart of some key stages in the expansion of our universe.

energies that all the particles in the Big Bang had during the first 10^{-14} seconds. Many crucial features of our universe could have been imprinted when the cosmic clock was reading 10^{-35} seconds, or even less. In these contexts, each factor of ten on the cosmic clock in the age of the universe – each extra zero after the decimal point – is likely to be equally eventful and should count equally. The leap back from 10^{-14} seconds to 10^{-35} seconds is then bigger (in that it spans more factors of ten) than the timespan between the three minute threshold when helium was formed (about 200 seconds after the Big Bang) and the present time (3×10^{17} seconds, or ten billion years). In this perspective, there is plenty of action at even earlier stages.

UNIFICATION IN THE MICROWORLD

Right back at the beginning, the mysteries of the cosmos and the microworld overlapped. To probe these mysteries, we need to relate gravity, the dominant force on large scales, to the other forces that govern individual particles. This is still unfinished business. But the various forces and particles of the subatomic world are now seen to fall into a pattern.

Early in the nineteenth century, Michael Faraday realized that electricity and magnetism were intimately linked: a moving magnet generated electric currents; a moving electric charge, conversely, created a magnetic field. This principle underlies electric motors and dynamos. In 1864 James Clark Maxwell codified Faraday's discoveries into a famous set of equations, which expressed how a changing electric field generates a magnetic field, and vice versa. In empty space, these equations have solutions where the electric and magnetic fields oscillate. This is what light is: it's a wave of electric and magnetic energy (as are radio waves, X-rays, and the rest of what we now call the electromagnetic spectrum).

This left just two distinct forces: electromagnetism (perceived as a single force) and gravity. Even Faraday yearned for a unification between gravity and electromagnetism, although he realized that it was premature. A hundred years on, Einstein spent his later years seeking a deep connection between these two forces. This was still a vain quest. Indeed, we now realize that it was doomed because he didn't then know about the short-range forces that govern atomic nuclei: the 'strong' or nuclear force that binds the protons and neutrons together in atomic nuclei (and determines our number ε); and the 'weak' force, important for radiative decay and neutrinos. In the somewhat harsh view of his most distinguished biographer, the physicist Abraham Pais, Einstein 'might as well have gone fishing' for the last thirty years of his life.

The challenge is now to unify *four* forces: the three that govern the microworld – electromagnetism, the nuclear force, and the 'weak' force – and the force of gravity. The first modern step towards this unification was associated with the names of Sheldon Glashow and Steven Weinberg in the US, Gerard t'Hooft in Holland, and the Pakistani physicist Abdus Salam. The outcome of their work was to show that the electric and magnetic forces (unified by Maxwell) are themselves linked to an apparently quite different force – the so-called 'weak' force important for neutrinos and radioactivity. These forces would have been the same in the very early universe; they acquired distinctive identities only after the universe had cooled below a critical temperature of about 10^{15} degrees (which happened when it was 10^{-12} seconds old). The biggest accelerators can simulate these temperatures, and Salam and Weinberg were vindicated when experiments at CERN discovered new particles that they had predicted.

In the 1950s and 1960s, so many new kinds of particles were discovered (supplementing the familiar electrons, neutrons and protons) that there seemed a risk that particle physics would become like stamp collecting. But patterns were discerned; the subatomic particles could be grouped into 'families', rather as the atoms in the periodic table fall into 'periods' and 'groups'. In 1964, Murray Gell-Mann and George Zweig, two American theorists, introduced the 'quark model'. Quarks have charges that are 1/3 or 2/3 that of the electron. Experimental support came from Jerome Friedman, Henry Kendall and Richard Taylor, who used the newly commissioned Stanford Linear Accelerator to crash electrons into protons. They found that the electrons scattered as though each proton was made up of three 'point charges', carrying respectively 2/3, 2/3 and −1/3 of the total charge. One counterintuitive aspect of the 'quark model', however, is that an isolated quark can never be dislodged even though, inside a proton, the quarks behave as though they are free. (All attempts to detect fractionally charged particles have failed.) By the late 1970s, most of the 'particle zoo' had been explained in terms of nine types of quark.

The so-called 'standard model' that emerged in the 1970s has brought impressive order into the microworld. The electromagnetic and 'weak' forces have been unified; and the strong or nuclear forces have been interpreted in terms of quarks, held together by another kind of particle called a 'gluon'. But nobody has taken this as the final word: the number of elementary particles remains bewilderingly large, and the equations still involve numbers that have to be determined by experiment and can't be derived from theory alone. In particular, the 'gluon' interpretation does not pin down the strength of nuclear forces, crucially manifested in our basic number $\varepsilon = 0.007$.

The next goal after unifying the electromagnetic and weak forces is to bring in the nuclear force, and thereby achieve a so-called 'grand unified theory' (GUT) of all the forces governing the microphysical world (although these theories are still not grand enough to include gravity, which poses a still greater challenge). A stumbling block is that the grand unification is thought to occur at a temperature of 10^{28} degrees. This is a million million times higher than experiments can presently reach – and to achieve the requisite energies would need an accelerator far bigger than our Solar System. It is hard, therefore, to test these theories on Earth.

Their distinctive consequences in our low-energy world are vestigial: for instance, protons, the main ingredient of all stars and planets, would very slowly decay – an effect that could be important in the remote future but is insignificant now. *Everything*, however, would have been hotter than 10^{28} degrees for the first 10^{-35} seconds. Perhaps the early universe was the only place where the requisite temperature for unifying the forces could even be reached. This 'experiment' shut down more than ten billion years ago, but did it leave fossils behind, just as most of the helium in the universe survives from the first few minutes? It seems that it did: indeed, the favouritism of matter over antimatter (discussed in Chapter 6) may have been imprinted at this ultra-early stage. Even more important, the vast scale of the universe, and the fact that it is expanding at all, may be determined by what happened in those brief initial instants.

THE 'INFLATION' CONCEPT

Two fundamental questions about our universe are: 'Why is it expanding?' and 'Why is it so big?' We can trace out what happens during the expansion, and we can extrapolate right back to the first few seconds (and corroborate this with the helium and deuterium abundance). But the so-called Big Bang theory is really a description (and a quite successful one) of what happened *after* the Big Bang. It says nothing about what set up the expansion in the first place. Another puzzle is: 'Why does our universe have the overall uniformity that makes cosmology tractable, while nonetheless allowing the formation of galaxies, clusters and superclusters?' And, still further: 'What imprinted the physical laws themselves?'

One basic mystery (discussed in Chapter 6) is why our universe is expanding, after ten billion years, with Ω still not too different from a value of one. Our universe has neither collapsed long ago, nor is it expanding so fast that its kinetic energy has overwhelmed the effect of gravity by many powers of ten. This requires Ω to have been tuned amazingly close to a value of unity in the early universe. What made everything start expanding in this special way? Why, when we observe remote regions in opposite directions, do they look so similar? Or why is the temperature of the microwave afterglow almost the same all over the sky?

These mysteries would be solved if all parts of our present universe had synchronized and co-ordinated themselves very early on, and then accelerated apart – and this is the key postulate of the 'inflationary universe' theory. The (then) young American physicist Alan Guth put forward this idea in 1981. As so often happens in science, there were several precursors, especially the theories of Alex Starobinski and Andrei Linde in the Soviet Union and Katsumoto Sato in Japan, but Guth made the arguments clear enough to convince most of us that this was indeed a crucial insight. His book *The Inflationary Universe*[2] recounts the '*eureka* moment' when the idea dawned on him, and how a lively community of theorists debated and developed it further. (Guth also offers frank sociological insight into the American academic scene, from the perspective of a young researcher seeking a niche in an overcrowded profession.)

According to the 'inflationary universe' theory, the reason why our universe is so big, and why gravity and expansion are so closely balanced, lies in something remarkable that happened *very* early on, when our entire observable universe was literally of microscopic size. At the colossal densities that then prevailed, a 'cosmic repulsion', rather like an enormously strong λ, came into play and overwhelmed ordinary gravity. The expansion was 'kicked into overdrive', leading to runaway acceleration, so that an embryo universe could have inflated, homogenized, and established the 'fine-tuned' balance between gravitational and kinetic energy.

All this is supposed to have happened within about 10^{-35}

seconds of the Big Bang! The conditions that prevailed back then are far beyond what we can test experimentally, and the details are therefore speculative. We can nonetheless make guesses consistent with other physical theories and with what we know about the later universe.

The idea behind the 'inflation' theory is compellingly attractive because it seems to show how an entire universe could evolve from a tiny 'seed'. This is deemed to have happened because the expansion is *exponential*; it doubles, then doubles, and then doubles again . . . Mathematical formulae (unless they are very long and complicated indeed) generally don't yield huge numbers. The only natural way for a 'modest' number to generate a gigantic one – such as 10^{78} , the total number of atoms in our observable universe – is if it is 'in the exponent' (to use mathematical jargon), so that it tells how many times the size doubles. Each time a sphere doubles its radius, its volume goes up by a factor of eight (in ordinary Euclidean space); only a hundred of these doublings would be needed in order to reach a number like 10^{78}.

This is just what is proposed as happening during the 'inflationary' phase of our universe. The fierce repulsion that drove inflation must have switched off, allowing the universe, having by then enlarged enough to encompass everything that we now see, to embark on its more leisurely expansion. This transition converted the huge energy latent in the original 'vacuum' into ordinary energy, generating the heat of the fireball and initiating the more familiar expansion process that has led to our present universe.

The concept of inflation has been boisterously debated ever since it was first proposed twenty years ago. It has been through many variants, based on different assumptions about how the pressure, density and so forth behaved under conditions far beyond anything that we can study directly. But the general idea will surely retain its appeal unless a better one comes along. At the moment, if offers the only credible explanation for why our universe is so large and so uniform. It suggests *why* the universe is expanding at such a seemingly fine-tuned rate, so that it could heave itself up to dimensions of ten billion light-years.

CAN WE TEST THE INFLATION THEORY?

If a wrinkled surface is stretched by a huge factor, then the curvature reduces until any deviations from flatness are imperceptible. The analogue of 'flatness' in cosmology is an exact balance between (negative) gravitational energy and (positive) expansion energy. This is the firmest generic prediction of inflation. Is it fulfilled? The simplest kind of flat universe is one in which Ω is exactly unity. The evidence in Chapter 5 that atoms and dark matter contribute only 0.3 of the critical density seemed at first sight to be a setback. Theorists therefore seized enthusiastically on the claim that the expansion is accelerating, because the energy associated with the number λ must then be added in. Our universe seems indeed to be 'flat' (though the more cautious among us may say the jury is still out, and await a definitive verdict within a few years). The 'mix' of stuff that makes up the critical density is four per cent atoms and about 25 per cent dark matter, the rest is the 'vacuum' itself.

This evidence of 'flatness' is moderately encouraging. It at least motivates us to seek further tests, especially 'diagnostics' that might reveal details of what happened during inflation. Most detailed ideas about the ultra-early universe have a short shelf-life. The first 10^{-35} seconds is as uncertain today as was the physics of one second after the Big Bang when Gamow and other pioneers first explored the cosmological origin of the elements. Their first ideas were wrong in important respects, but were corrected and put on a firm footing within a decade or two. Maybe we can share similar hopes about a symbiosis between ultra-high-energy physics and cosmology in the next decade.

Helium formation in the first few minutes involved nuclear

reactions and atomic collisions of a kind that can be reproduced experimentally. In contrast, the processes during the inflationary era that determine fundamental cosmic numbers such as Q are too extreme to be simulated terrestrially, even in accelerators. That makes the new challenge more daunting. On the other hand, that very fact provides an extra motive for studying the very early universe. It may offer the firmest tests of new unified theories because it is the *only place* where energies are high enough for the distinctive consequences of these theories to be manifested. When astronomers are trying to understand cosmic phenomena, they normally utilize discoveries made by physicists in the lab. Perhaps they can now return the compliment by discovering some fundamentally new physics. There are already other instances of this – for instance, neutron stars extend our knowledge of dense matter and strong gravity. But most extreme of all is the Big Bang itself. In the 1950s, cosmology was outside the mainstream of physics – only a few 'eccentrics' like Gamow paid any attention to it. In contrast, cosmological issues now engage the interest of many leading mainstream theoretical physicists. And that surely gives us grounds for optimism.

Microscopic 'vibrations', imprinted when our universe was smaller than a golfball, inflate so much that they now stretch across the universe, constituting the ripples that develop into galaxies and clusters of galaxies. Theorists still haven't shown whether inflationary models can 'naturally' account for $Q = 10^{-5}$ characterizing the amplitude of these ripples; it depends on some physics that is still anything but 'battle-tested'. But we can learn something about the details (and rule out some options) because specific variants of inflation make distinctive predictions. Measurements with the MAP and Planck-Surveyor spacecraft, and surveys of how galaxies are clustered, will offer clues about the inflationary phase, and teach us things about 'grand unified' physics that can't be directly inferred from experiments at 'ordinary' energy levels.

Along with the fluctuations that develop into galaxies and

clusters, the inflation is thought to generate 'gravitational waves' – oscillations in the fabric of space itself, criss-crossing the universe at the speed of light. Objects encountered by such waves feel a gravitational force that pulls them first one way and then the other; they 'shake' slightly as a result. The effect is minuscule and its detection in reality poses a formidable technical challenge. The European Space Agency's LISA project (standing for Laser Interferometric Space Array) is planned to deploy a set of spacecraft in orbits around the Sun, separated by several million kilometres. The distances between them would be monitored by laser beams to a precision of a millionth of a metre.

Even LISA may not prove sensitive enough to 'feel' these primordial vibrations. It is therefore a comfort to its designers that other signals should be easier to detect. An intense burst of gravitational waves would, for instance, be generated whenever two black holes collided and coalesced. We expect such events to occur from time to time. Most galaxies harbour a central hole as massive as millions of stars. Pairs of galaxies often collide and merge (we see many such events in progress); whenever this happens, the holes in the centres of the two participating galaxies spiral together.

We can therefore look forward soon to empirical probes of the inflation era. Even if we don't know the appropriate physics, we can calculate the quantitative consequences of specific assumptions of the theory (the value of Q, the gravitational waves, etc). We can then compare these with the observations, and thereby at least constrain the possibilities.

OTHER RELICS

Any 'fossils' of that ultra-early era would be important as missing links between the cosmos and the microworld. One interesting possibility (which loomed large in Guth's mind when he was developing his theory) is that magnetic monopoles might have survived from the early universe. Faraday and Maxwell showed the intimate relation between electricity and magnetism, but there was (as they well realized) one key difference between these two forces: positive and negative electric charges exist, but 'north' and 'south' magnetic poles don't seem to come separately. Magnets are dipoles (with two poles) rather than monopoles (with one); and if we chop up a dipole we never get two monopoles, merely smaller dipoles. Despite many ingenious searches, nobody has ever 'caught' a monopole.

Modern theories suggest that monopoles could exist, but they may be immensely heavy (a million billion times heavier than a proton). Because of the high mass, it would need an immense concentration of energy to make them – the kind of energies that prevailed in the very early universe but not thereafter. There are very few monopoles in our present universe – magnetic fields pervade interstellar space, and these would be 'shorted out' if there were a population of monopoles. Guth was puzzled by the absence of monopoles because it seemed that they would unavoidably have been produced in the early universe – indeed, his best guess was that their total collective mass would amount to millions of times more dark matter than there actually is. An important bonus of inflation (if it occurred after the monopoles formed) is that it would dilute the putative monopoles, and thereby account for their apparent absence today.

Monopoles are a kind of 'knot' in space – in the jargon of the subject, they are 'topological defects'. Even more interesting are defects in the form of lines rather than points – regions of space that get knotted into tubes far thinner than an atom. They would either make closed loops, like elastic bands, flailing around at nearly the speed of light, or else stretch right across the universe. Some cosmologists have speculated that these defects in space could be the seeds for cosmic structure – in effect, that they contribute to Q. This idea attracted interest in the early 1990s, but turned out to be incompatible with the details of the galaxy clustering that was subsequently mapped out. But these loops could still exist, and they are so extraordinary in their properties (thinner than an atom, but so heavy that each kilometre could weigh as much as the Earth) that astronomers should make every effort to find one.

Miniature black holes are another interesting possibility. A hole the size of a single atom would be as massive as a mountain. As we've seen in Chapter 3, this is a direct result of $\mathcal{N}$ being so large: gravity is so feeble that it can't overwhelm other forces on the atomic scale unless the mass of $\mathcal{N}$ atoms is packed into the volume of one. Conceivably the ultra-early universe generated the requisite pressures to make them. Even though no present-day process could provide this degree of implosion, maybe some future high-tech civilization could do so – an especially fascinating prospect if combined with the other speculation that, within a black hole, a new universe may sprout and inflate into a new (possibly infinite) space–time disconnected from ours.

FROM 'NOTHING'?

It may seem counterintuitive that an entire universe ten billion light-years across (and which probably spreads even further beyond our horizon) can have emerged from an infinitesimal speck. What makes this possible is that, however much inflation has occurred, the universe's net energy can still be zero. Everything has energy mc^2, according to Einstein's famous equation. But everything also has negative energy because of gravity. We need energy to escape from Earth's gravity – the burning of enough rocket fuel to reach a speed of 11.2 kilometres per second. Down on the Earth's surface we therefore have an energy deficit compared with an astronaut in space. But the deficit (technically called 'gravitational potential energy') due to everything in the universe added together could amount to *minus* mc^2. In other words, the universe makes for itself a 'gravitational pit' so deep that everything in it has a negative gravitational energy that exactly compensates for its rest-mass energy. So the energy cost of inflating our universe could actually be zero.

Cosmologists sometimes claim that the universe can arise 'from nothing'. But they should watch their language, especially when addressing philosophers. We've realized ever since Einstein that empty space can have a structure such that it can be warped and distorted. Even if shrunk to a 'point', it is latent with particles and forces – still a far richer construct than the philosopher's 'nothing'. Theorists may, some day, be able to write down fundamental equations governing physical reality. But physics can never explain what 'breathes fire' into the equations, and actualizes them in a real cosmos. The fundamental question of 'Why is there something rather than nothing?' remains the province of philosophers. And even they may be wiser to respond, with Ludwig Wittgenstein, that 'whereof one cannot speak, one must be silent'.

BEYOND OUR HORIZON TO THE MULTIVERSE

The long-range forecasts sketched in Chapter 7 were actually based on an assumption that we can't test, namely that the parts of the universe beyond our present horizon resemble those we see. If you were in the middle of an ocean, you wouldn't expect land to lie immediately over the horizon; but you'd know that the ocean wasn't unending and would eventually be bounded by a continent. Likewise, we may be mistaken in thinking that our universe extends uniformly without limit. We could perhaps be living in a low-density bubble, big enough that its edge lies far beyond our present horizon, yet surrounded by a still larger region that will eventually collapse on top of us. If so, our remote descendants would revise the 'forecast' of perpetual expansion when the higher-density material loomed within their horizon. A drastic change only just beyond our horizon would be unlikely; on the other hand, we have no warrant to extrapolate all the way to infinity.

The most important implication of inflation is that it grandly and dramatically enlarges our perspective on the universe. To explain the universe that we see, there must have been enough inflation to account for the 10^{78} atoms within range of our telescopes. But that's just a minimum. It may take a long time to stop the inflation once it has started (theorists refer to this as the problem of the 'graceful exit' from inflation). Indeed, most versions of the theory suggest that the number of 'doublings' should be *far more* than is needed to account for our observable universe. In Chapter 1, we imagined a succession of views of our universe, each taken ten times further away than the last. Twenty-five frames took us to the limit of our present vision, starting from the everyday human scale. This limit is set, essentially, by how far light has been able to travel in the ten billion years or so since the first galaxies formed. But inflation theorists envisage a universe so much larger that it would take *millions* of frames, each a leap by a factor of ten, to reach any 'edge'. This stupendous expanse of space is (to me at least) impossible to grasp. The leap in scale from the microworld to our horizon is as nothing compared with the leap beyond that to the real limit of our universe. Though not infinite, our domain of space and time extends far beyond what we can see. The time before light reaches us from the 'edge' is then a number of years written not just within ten zeros, nor even with a hundred, but with millions.

But this isn't all. Even this colossal universe, whose extent requires a million-digit number to express it, may not be 'everything there is'. It is the outcome of one episode of inflation; but that episode – that Big Bang – may itself just be one event in an infinite ensemble. Indeed, this is a natural consequence of the 'eternal inflation' espoused especially by the Russian cosmologist Andrei Linde. According to this scenario, which requires specific (though still speculative) assumptions about the physics at extreme densities, the cosmos may have had an infinite past. Patches where inflation doesn't end always grow fast enough to provide the seeds for other Big Bangs. There are variants on these speculations, in which an episode of inflation could be triggered inside a black hole, creating new domains of space and time disjoint from our own.

At this point, let me add a semantic note about the definition of 'universe'. The proper definition of 'universe' is, of course, 'everything there is'. I am arguing in this chapter that the entity traditionally called 'the universe' – what astronomers study, or the aftermath of our Big Bang – may be just one of a whole ensemble, each one maybe starting with its own Big Bang. Pedants might prefer to redefine the whole ensemble as 'the universe'. But I think it is less confusing to leave the term 'universe' for what it has traditionally connoted, even though this then demands a new word, the 'multiverse', for the entire ensemble of 'universes' – a concept to which I'll return in Chapter 11.

CHAPTER 10

THREE DIMENSIONS (AND MORE)

> The Earth's orbit is the measure of all things; circumscribe around it a dodecahedron and the circle containing this will be Mars; circumscribe around Mars a tetrahedron, and a circle containing this will be Jupiter; circumscribe around Jupiter a cube, and the circle containing this will be Saturn. Now inscribe within the Earth an icosahedron, and the circle contained in it will be Venus; inscribe within Venus an octahedron, and the circle contained in it will be Mercury. You now have the reason for the number of the planets.
>
> Johannes Kepler

WHY $\mathcal{D}$ = 3 IS SPECIAL

Our space has three dimensions. There are points (zero dimensions), lines (one dimension), surfaces (two dimensions) and solid objects (three dimensions). But there the sequence stops, even though mathematically we can imagine a kind of space that has more. What is special about the number three? From classical times, geometers have noted interesting features of different dimensions. For example, in two dimensions we can draw a regular polygon with any number of equal sides (an equilateral triangle, a square, a pentagon, a hexagon, etc). But in three dimensions there are just the five Platonic 'regular solids', in which all sides and all angles are equal. In four dimensions there are six such objects, and in all higher dimensions there are just three.

One consequence of a three-dimensional world is that forces like gravity and electricity obey an inverse-square law, such that the force from a mass or charge is four times weaker if you go twice as far away. Michael Faraday, in his pioneering studies of electricity, had a graphic (and essentially correct) way to understand this. He envisaged 'lines of force' sprouting from every charge or mass, the strength of the force depending on how concentrated the lines are. At a distance r, the lines are spread out over an area proportional to r^2; at larger distances, the force is consequently diluted, its strength depending inversely on r^2. However, the area of a *four*-dimensional 'sphere' would vary in proportion to r^3 – it would be eight, not just four, times larger if r doubled in value. Faraday's argument would then imply an inverse-*cube* law.

As Newton realized, the trajectories of planets are controlled by a balance between the effects of gravity, tending to pull them inward, and the centrifugal effect of their motion. Orbits in our Solar System are *stable*, in the sense that a slight change in a planet's speed would only nudge its orbit slightly. But this stability would be lost if gravity followed an inverse-cube (or steeper) law rather than one based on inverse squares. An orbiting planet that was slowed down – even slightly – would then plunge ever-faster into the Sun, rather than merely shift into a slightly smaller orbit, because an inverse-cube force strengthens so steeply towards the centre; conversely, an orbiting planet that was slightly speeded up would quickly spiral outwards into darkness.

The eighteenth-century English theologian William Paley is famous for his argument that the apparent design in our universe implies a Designer, just as a watch implies a watchmaker. Paley had been well-enough trained in mathematics at Cambridge to appreciate this arcane feature of the inverse-square law, and included it in his armoury of argument for a benign Creator. Most of his other 'evidences of design' came from biology and have since been discounted, even by theologians, in the post-Darwinian era. The impressive adaptation of the eye, the hand and so forth, are the outcome of natural selection, and of symbiosis between living organisms and their environment. Paley's argument that the inverse-square law is especially benign now seems one of his more robust ones: there is no scope for natural selection of a favoured law of force, and nothing could react back on the universe to change it. Paley was writing more than a century before atoms were realized to consist of electrons orbiting a positively charged nucleus; otherwise, he could have bolstered his case by noting that, for similar reasons, atoms would be impossible in a universe ruled by an inverse-cube law because there would be no stable orbits for electrons.

There is therefore a problem with more than three spatial dimensions. Could we then live in a world where there were *less* than three? The best argument here is a very simple one: there are inherent limitations on complex structures in 'flatland' (or, indeed, on any two-dimensional surface). It is impossible to have a complicated network without the wires crossing; nor can an object have a channel through it (a digestive tract, for instance) without dividing into two. And the scope is still more constricted in a one-dimensional 'lineland'.

These are just the most obvious reasons – and mathematicians have discovered others – as to why we shouldn't be surprised that we find ourselves living in three-dimensional space.

TIME AND ITS ARROW

Time is, of course, a fourth dimension that we experience. To locate an event, we need four numbers: three spatial coordinates to describe *where* it happened, and a fourth to tell us *when*. As an anonymous graffiti-writer put it: 'Time is nature's way of stopping things happening all at once'. Events are, as it were, strung out along paths whose milestones are the ticks of clocks. But time is different from the other three dimensions insofar as we seem to be dragged only one way in it ('forward'); in the other three, we can move in either direction (East or West, North or South, up or down). Our universe is thus best described as (3 + 1)-dimensional. Einstein taught us that space and time are linked, and that the rate at which time passes is 'elastic', being dependent on how a timekeeper is moving, and whether or not that timekeeper is near to a large mass. But Einstein's ideas retain a distinction between time and space – between what's out there in space and what lies in the past or future.

An 'arrow of time' points insistently from the past towards the future. A film of everyday events looks grotesquely different when run backwards. Cause and effect are reversed; broken bits of glass, and drops of liquid, seem to rush purposefully together to assemble into a glassful of wine; steam converging on a kettle condenses into water. In Martin Amis's ironic time-reversed novel *Time's Arrow*[1] New York taxi cabs 'pay you up front, no questions asked . . . no wonder we stand there, for hours on end, waving goodbye, or saluting – saluting this fine service'.

The asymmetry between past and future is so ingrained in our experience that few, except for some philosophical physicists, pause to ponder the conundrums it poses. It is perplexing because no such asymmetry is built into the basic laws governing the microworld. The world changes irreversibly, even though the underlying laws are indifferent between past and future. A film showing a single collision between two snooker balls would look more or less the same whether it was run backwards or forwards, but the whole pattern of collisions after play begins plainly displays an arrow of time. Likewise, our world seems to have been set up in a special way.

We are 'trapped' in time; but we could achieve clearer insights from an imaginary perspective that is 'outside time' –

like the creatures in Kurt Vonnegut's *Sirens of Titan*, who perceive people as 'great millipedes with babies' legs at one end and old people's legs at the other'. Our universe would then appear as a static four-dimensional entity (the 'block universe'); the 'world lines' of everyday objects would then be more disordered at one end (what we call the future) than at the other end (what we call the past). But the hard thing to explain is *any* 'ordered' state at all. If one end of a long piece of string is woven into a remarkable pattern, we are equally surprised, whether it is the left-hand or right-hand end. Likewise, in a 'block universe', where the future seems to exist on the same footing as the past, it is no more (and no less) puzzling to find order at the start than at the finish.

When we say the universe is expanding we are of course presupposing an arrow of time, and that we can order the frames in a movie (or the three-dimensional slices in our 'block universe') so that the universe is more dispersed at times we designate as 'later'.

The asymmetry in time may be linked to the expansion of the universe. Indeed, I have described in Chapter 8 how, during the expansion, gravity enhances any initial density contrasts, allowing structure to emerge from a fireball that started off almost featureless. In the early stages, this asymmetry wouldn't show up in any local measurement, because the density would at that time be so high that microscopic processes – collisions between particles, the emission and absorption of photons, etc – would occur very fast compared with the expansion rate. Everything would, at every instant, be in equilibrium. The material would retain no 'memory' of whether it had previously been denser or less dense, and would bear no imprint of the direction of time. But when the universe is more dilute, these reactions get slower, and the expansion then makes a crucial difference.

For example, if our universe had remained at a temperature of a billion degrees for a long time, or if nuclear reactions had happened faster, all the atoms would have been processed into iron. Fortunately, the expansion was fast enough to quench nuclear reactions before they could do more than convert twenty-three per cent of the hydrogen into helium. This exemplifies how cosmic expansion allows departures from equilibrium, so that what happens is not the same as would have been in a contracting universe.

As Sakharov first pointed out, our very existence depends on an irreversible effect that established an excess of matter over antimatter at a still earlier stage. Had that not occurred, all the matter would have been annihilated with an equal amount of antimatter, leaving a universe containing no atoms at all. There would then have been no stars, still less any of the chemistry that has allowed complex structures to emerge.

Time still poses mysteries on which there is absolutely no consensus. The physicist Julian Barbour conducted an informal poll among experts on the question 'Do you believe time is a truly basic concept . . . [or can it] be derived from more primitive notions (rather as, for instance, an object's temperature derives from the agitated motions of its consistent atoms)?' Responses were quite evenly divided, with a slight majority favouring the view that time would eventually be interpreted in terms of something deeper.

WRAPPED-UP DIMENSIONS ON LARGE SCALES?

Space and time certainly have a complicated structure. We know that space is punctured by black holes – millions within our galaxy, even bigger ones in the centres of other galaxies – in which time and space are intertwined. But these complications are restricted to regions that are 'local' in a cosmological perspective. The near-uniformity of our universe on scales larger than superclusters suggests that the geometry of space is smooth and simple on the scale of our present horizon. So also does the fact that the background microwave radiation has almost the same temperature over the whole sky.

Mathematically inclined cosmologists have nonetheless

wondered whether this simplicity could be an illusion: perhaps we actually see the same patch over and over again, as in a hall of mirrors or a kaleidoscope – space being 'rolled up' or having some kind of cellular structure. If we were indeed in this strange kind of universe, the cells must be at least a few per cent of our horizon distance (in other words, more than a few hundred million light-years across): we know this because, if the cells were smaller, we would see distinctive structures like the Virgo cluster of galaxies repeating themselves. A stronger constraint has now come from measurements of small non-uniformities in the microwave background temperature over the sky. There is no repetitive pattern in these non-uniformities, and so we can now rule out any cell size much smaller than our horizon.

Beyond the horizon set by the finite speed of light, observations tell us little. Space could be wrapped in a complicated way on scales far exceeding ten billion light-years. There could even be changes in the number of dimensions. But we shall never have more than indirect intimations of what happens beyond the range of any telescope.

What about ultra-*small* scales? Here, our simple concepts certainly break down. Indeed, we may need to grapple with very complicated ideas, involving extra dimensions, in order properly to understand the particles, the forces, and our cosmic numbers.

THE MICROSTRUCTURE OF SPACE AND TIME: QUANTUM GRAVITY

We have had a century to get used to the idea that ordinary materials – solids, liquids and gases – have a discrete atomic or molecular structure. Could there even be graininess in space and time themselves? Space seems a smooth continuum, but that is only because our experience, and even our most sophisticated experiments, are too 'coarse' to probe the very fine scale on which this structure would be manifest.

We don't know the detailed microstructure of space and time, but very general arguments tell us that it can't be chopped up into arbitrarily small pieces. Fine-scale detail can only be probed by radiation with wavelengths still shorter than that scale. For example, a building doesn't obstruct radio waves with wavelengths of many metres, but it casts sharp shadows in sunlight. Light consists of waves a millionth of a metre long, and nothing smaller than that can be imaged with an ordinary optical microscope: to probe sharper detail requires still shorter wavelengths (or else some other technique, such as an electron microscope). But, according to the quantum theory, shorter wavelengths come in more energetic quanta, or 'packets', of energy.

The basic quantum of energy is measured by Planck's constant (a number named after the great physicist Max Planck, who pioneered the idea of quantization a century ago). Up to a point, we can probe ever-finer detail by using more and more energetic quanta, associated with ever-shorter wavelengths. But there is a limit. This limit arises when the requisite quanta are such extreme concentrations of energy that they collapse into black holes. This happens at the 'Planck length', which is about 10^{19} times smaller than a proton; quanta with this tiny wavelength each carry as much energy as the rest-mass of 10^{19} protons. Light takes about 10^{-43} seconds to traverse this distance, and this 'Planck time' is the shortest time interval that can ever be measured. So even space and time are subject to quantum effects. However, because gravity is so weak, these effects come in on a far smaller scale than in ordinary atoms, when the controlling forces are electrical. (This is a consequence of the vastness of our first cosmic number, $\mathcal{N}$.)

Some theorists are more willing to speculate than others. But even the boldest acknowledge the 'Planck scales' as an ultimate barrier. We cannot measure distances smaller than the Planck length; we cannot distinguish two events (or decide which came first) when the time interval between them is less than the Planck time. These scales are smaller than atoms by just as much as atoms are smaller than stars. There is no prospect of any direct measurements in this domain: it would require particles with energies a million billion times higher than can be produced in the laboratory.

The two great 'pillars' of twentieth-century science are quantum mechanics, crucial in the microworld, and Einstein's theory of gravity, which does not incorporate quantum concepts. But we have no single framework that reconciles and unifies them. This lack doesn't impede the progress of terrestrial science, nor indeed the advance of astronomy, because most phenomena involve *either* quantum effects *or* gravity, but not both. Gravity is negligible, by our huge number $\mathcal{N}$, in the microworld of atoms or molecules, where quantum effects are crucial; conversely, quantum uncertainty can be ignored in the celestial realm of planets, stars and galaxies, where gravity holds sway. But right back at the beginning, quantum vibrations could shake the whole universe. Conversely, gravity could be important on the scale of a single quantum. This happens at 10^{-43} seconds, the Planck time. To understand the first instants after the Big Bang, or the space and time near the 'singularity' inside black holes, we need a unification of quantum theory and gravity.

Ordinary intuition breaks down at speeds approaching that of light, and near black holes. And it breaks down, too, at the extreme conditions of the very early universe, and on microscales close to the Planck length. We must then jettison cherished commonsense notions of space and time: black holes may be appearing and disappearing; space-time on this tiny scale may have a chaotic foam-like structure, with no well-defined arrow of time. The fluctuations may spawn new domains that evolve into separate universes. Space may have a kind of lattice structure, or be knotted rather like chain-mail. Time may become like space, so that in a sense there is no beginning of time.

The only other arena for quantum gravity is the central

singularity within black holes, shrouded within the horizon. A theory that has no manifest consequences except in such exotic and inaccessible domains is hard to check. To be taken seriously, it must either be rigidly embedded in some all-embracing theory that can be tested in many other ways or else it must be perceived to have a unique inevitability about it.

Several approaches are being followed, but there is no consensus yet about which is the right one. (Stephen Hawking now bets 'evens' that a unified theory will come within twenty years, although he admits he recently had to pay up after losing a similar bet that he made twenty years earlier!) The most ambitious and encouraging approach seems to be *superstring theory*, which leapfrogs directly to a unified theory of all the forces, and yields quantum gravity almost as a bonus.

SUPERSTRINGS

Superstring theory can, its proponents claim, incorporate the three forces that govern the microworld – electromagnetism, the nuclear force, and the 'weak' force – as well as accounting for the elementary particles (quarks, gluons, etc). The existence of gravity is actually an essential ingredient of the theory rather than an extra complication. Its key idea is that the fundamental entities in our universe are not points but tiny string loops, and that the various subnuclear particles are different modes of vibration – different harmonics – of these strings. The strings have the scale of the Planck length; in other words, they are many factors of ten smaller than we can actually probe. Moreover, these strings are vibrating not in our ordinary (3 + 1)-dimensional space, but in a space of ten dimensions.

The idea of extra dimensions is not a new one. Back in the 1920s Theodor Kaluza and Oskar Klein attempted to extend

Einstein's theory of space and time to include electrical forces. They tried to envisage electric fields, and the motions of charged particles by, as it were, attaching extra structure to each point in our ordinary space. The extra dimension was 'wound up' on a tiny scale, and didn't manifest itself to us, rather as a sheet of paper looks like a one-dimensional line when rolled very tightly, even though it is actually a two-dimensional surface. The Kaluza-Klein theory ran into difficulties, but the concept of extra dimensions has, more recently, had a dramatic renaissance. In superstring theory, each 'point' in our ordinary space is a complicated geometrical structure in *six* dimensions, wrapped up on the scale of the Planck length.

All physical theories involve equations and formulae that render the technicalities (though not, fortunately, the key ideas) opaque to the non-specialist. But, generally, the mathematics has already been worked out and can be taken 'off the shelf' by the physicists. For instance, the geometrical concepts that Einstein used in his theory of 'curved space-time' had all been developed in the nineteenth century; so also was the mathematical language that was deployed to describe the quantum world. But superstrings pose questions that still baffle mathematicians. For instance, is there any particular reason why a universe should end up with *four* 'expanded' dimensions (time, plus three dimensions of space), rather than some different number? The nature of our world, and the forces governing it, would depend on exactly how the extra dimensions 'wrapped up'. How does this come about, and are there a lot of different ways in which it could happen?

Superstring theories first aroused enthusiasm in the 1980s (even though the ideas go back to earlier decades), and they have absorbed the effort since then of whole cohorts of brilliant mathematical physicists. Initial over-exuberance was succeeded by a period of discouragement, because of the theory's bewildering complexity. But, since 1995, superstrings have had a 'second wind'. It has been realized that the extra dimensions can wrap up into just five distinct classes of six-dimensional space; at a still deeper mathematical level, these may be separate but related structures embedded in an eleven-dimensional space. Furthermore the concept of strings (one-dimensional entities) can be broadened to include two-dimensional surfaces (membranes); indeed, in ten-dimensional space there can be higher dimensional surfaces: in other words, if a two-dimensional surface is called a two-brane, there can also be three-branes, and so on. There is still, however, an unbridged gap between the intricate complexity of ten-dimensional string theory and any phenomena that we can observe or measure.

There are earlier precedents for theories being taken very seriously even without direct empirical support, particularly in cases where they seem to have a unique 'elegance' or 'rightness' – a resounding ring of truth that compels assent. Many physicists in the 1920s, for example, were receptive to Einstein's theory of general relativity because of its immense conceptual appeal. It is now confirmed by precise observations, but in the early days the evidence was sparse. Einstein was himself more impressed by his theory's elegance than by any experiments. Likewise, in the present era Edward Witten, the currently acknowledged intellectual leader of mathematical physics, has said that 'good wrong ideas are extremely scarce, and good wrong ideas that even remotely rival the majesty of string theory have never been seen'.

Nonetheless, there are specific non-aesthetic reasons for being optimistic about superstrings. The first is that Einstein's theory of general relativity, which interprets gravity as curvature in four-dimensional space-time, is inescapably built into superstring theory. The long-sought synthesis between gravity and the quantum principle should thus naturally emerge.

And already the theory has offered a deeper understanding of black holes. The story here dates back to the early 1970s. Jacob Bekenstein, an Israeli physicist working at Princeton University, was pondering the consequences of the then-recent discovery that black holes were standardized objects (as mentioned already in Chapter 3). This implied that they lost all memory of how they were formed. There appeared to be immense numbers of ways in which a black hole could be built up – rocks, planets, gas or even spaceships could in principle fall into it – but all trace of this history was seemingly erased. Bekenstein noted that this was like the 'entropy increase' that occurs when two gases mix: many possible initial states lead to indistinguishable final states. A loss of information corresponds to an increase of *entropy*, and Bekenstein conjectured that a black hole might have an entropy that was a measure of the number of different ways in which it could have formed. If Bekenstein were right, black holes would also have a *temperature*, and his idea was put on a much firmer footing when Hawking calculated that black holes were not absolutely black, but would actually emit radiation. (The emission is far too slight to be measurable in the black holes that astronomers have discovered, but could be important if the atom-sized 'miniholes' described in Chapter 3 are actually found to exist).

Superstring theories, which describe the structure of space on the Planck scale, offered a new insight. The US theorist Andrew Strominger showed in 1996 how black holes (albeit of a special kind) could be imagined as 'built up' from string-scale elements, and showed how to calculate the number of 'rearrangements' of these tiny building blocks that would lead to the same hole. It agreed precisely with the value of the entropy calculated by Bekenstein and Hawking. This is not, of course, an empirical argument; but it enhances our confidence in the theory by corroborating a calculation based on more traditional physics, and it deepens our insight into a mysterious feature of black holes.

Another hope – although this is at present more controversial and less firmly based – is that superstrings may offer new insights into the concepts of the quantum. Richard Feynman said that 'nobody really understands quantum mechanics'. It works marvellously; most scientists apply it almost unthinkingly; but it has its 'spooky' aspects, which many thinkers, from Einstein onwards, have found hard to stomach; and it's hard to believe that we've already attained the optimum perspective on it.

Even if we can't directly probe the Planck scale, some features of the physical world that we *do* observe – for instance, the contingency that there are three basic forces in the microworld, particular types of particle, and so forth – may 'pop out' of superstring theory, just as Einstein's theory of gravity seems to. We would certainly gain confidence in the entire mathematical construct if this happened. Superstring theory may, as discussed in the next chapter, offer an overarching theory of the multiverse.

CHAPTER 11

COINCIDENCE, PROVIDENCE – OR MULTIVERSE?

On religion I tend towards deism but consider its proof largely a problem in astrophysics. The existence of a cosmological God who created the universe (as envisaged by deism) is possible, and may eventually be settled, perhaps by forms of material evidence not yet imagined.

E. O. Wilson, *Consilence*

WHAT DOES THE FINE TUNING MEAN?

In our universe, intricate complexity has unfolded from simple laws. But it's not guaranteed that simple laws permit complex consequences; indeed, we've seen that different choices of our six numbers would yield a boring or sterile universe. Similarly, mathematical formulae can have very rich implications, but generally they don't. The Mandelbrot set, for instance, with its infinite depth of intricate structure, is encoded by a short algorithm (see Figure 11.1). But other algorithms, superficially similar, yield very dull patterns.

There are various ways of reacting to the apparent fine tuning of our six numbers. One hard-headed response is that we couldn't exist if these numbers weren't adjusted in the appropriate 'special' way: we manifestly *are* here, so there's nothing to be surprised about. Many scientists take this line, but it certainly leaves *me* unsatisfied. I'm impressed by a

FIGURE 11.1

The Mandelbrot set. This infinitely complex pattern, which contains layer upon layer of intricate structure, is encoded by a short and simple algorithm. But many similar-seeming algorithms describe dull and featureless patterns. Our universe is governed by laws that permit immensely varied consequences.

metaphor given by the Canadian philosopher John Leslie. Suppose you are facing a firing squad. Fifty marksmen take aim, but they all miss. If they hadn't all missed, you wouldn't have survived to ponder the matter. But you wouldn't just leave it at that – you'd still be baffled, and would seek some further reason for your good fortune.

Others adduce the 'tuning' of the numbers as evidence for a beneficent Creator, who formed the universe with the specific intention of producing us (or, less anthropocentrically, of permitting intricate complexities to unfold). This is in the tradition of William Paley and other advocates of the so-called 'argument from design' for God's existence. Variants of it are now espoused by eminent scientist-theologians such as John Polkinghorne; he writes that the universe is 'not just "any old world", but it's special and finely tuned for life because it is the creation of a Creator who wills that it should be so'.[1]

If one doesn't accept the 'providence' argument, there is another perspective, which – though still conjectural – I find compellingly attractive. It is that our Big Bang may not have been the only one. Separate universes may have cooled down differently, ending up governed by different laws and defined by different numbers. This may not seem an 'economical' hypothesis – indeed, nothing might seem more extravagant than invoking multiple universes – but it is a natural deduction from some (albeit speculative) theories, and opens up a new vision of our universe as just one 'atom' selected from an infinite multiverse.

THE MULTIVERSE

Some people may be inclined to dismiss such concepts as 'metaphysics' (a damning put-down from a physicist's viewpoint). But I think the multiverse genuinely lies within the province of science, even though it is plainly still no more than a tentative hypothesis. This is because we can already map out what questions must be addressed in order to put it on a more credible footing; more importantly (since any good scientific theory must be vulnerable to being refuted), we can envisage some developments that might rule out the concept.

The prime stumbling-block is, of course, our perplexity about the extreme physics that applied in the initial instants after the Big Bang. There are strengthening reasons to take 'inflation' seriously as an explanation for our expanding universe: the theory's firmest and most generic prediction, that the universe should be 'flat', is seemingly borne out by the latest data (albeit not in the simplest form: three ingredients – atoms, dark matter, and the vacuum energy λ – contribute to the 'flatness'). The actual details of inflation depend on the physical laws that prevailed in the first 10^{-35}seconds, when conditions were so extreme as to be far beyond the range of direct experiment. But there are two ways we can, realistically, hope to pin down what those conditions were. Firstly, the ultra-early universe may have left conspicuous 'fossils' in our present-day universe. For example, clusters and superclusters of galaxies were 'seeded' by microscopic fluctuations that arose during inflation, and their detailed properties, which astronomers can now study, hold clues to the exotic physics that prevailed when these structures were laid down. Secondly, a unified theory may earn credibility by offering new insight into aspects of the microworld that now seem arbitrary and mysterious – for instance, the various types of subatomic particles (quarks, gluons, and so forth) and how they behave. We would then have confidence in applying the theory to the inflationary era.

Advances along these two routes may disclose to us a convincing description of the physics of the ultra-early universe. Computer simulations of how universes emerge from something of microscopic size would then be just as believable as our current calculations of how helium and deuterium were formed in the first few minutes of the expansion (Chapter 5) and how galaxies and clusters emerged from small fluctuations (Chapter 8).

Linde and others have already simulated some 'virtual multiverses', but at the time of writing this book the input into their calculations is highly arbitrary: many speculative options seem open, and we have no way of deciding among them. These studies of 'eternal inflation' (described in Chapter 9) already show us that some sets of assumptions, consistent with everything else we know, yield many universes that sprout from separate Big Bangs into disjoint regions of space-time. These universes would never be directly observable, even in principle; we couldn't even meaningfully say whether they existed 'before', 'after' or 'alongside' our own. However, if the input theory that predicted multiple universes could be 'battle-tested' by convincingly explaining things we *could* observe, then we should take the other (unobservable) universes seriously, just as we give credence to what our current theories predict about quarks inside atoms, or the regions shrouded inside black holes.

If there are indeed many universes, the next question that arises is: How much variety do they display? The answer again depends on the character of the physical laws at a deeper and more unified level than we yet understand. Perhaps some 'final theory' will give unique formulae for all of our six numbers. If it were to, then the other universes, even if they existed, would in essence be just replicas of ours, and the apparent 'tuning' would be no less a mystery than if our single universe were the whole of reality. We'd still be perplexed that a set of numbers imprinted in the extreme conditions of the Big Bang happened to lie in the narrow range that allowed such interesting consequences ten billion years later.

But there's another possibility. The underlying laws that apply throughout the multiverse may turn out to be more permissive. Each universe may evolve in a distinctive way, being characterized by a different set of numbers from those that are so crucial moulding our own universe. We are used to explaining contingencies here on the Earth (why there is a particular mountain, for instance), and even features in space (the shape of a nebula, the pattern of the galaxies), as 'accidents of history'. We can't explain such things any more deeply, although we don't doubt that they are the outcome of some underlying laws. By extension, the strength of the forces and the masses of elementary particles (as well as Ω, Q and λ) could be secondary outcomes of the final theory (maybe a version of superstring theory) that governs the entire multi-verse.

There is an analogy here with a 'phase transition', such as the familiar phenomenon of water turning into ice. When the inflationary era of a particular universe ended, space itself (the 'vacuum') underwent a drastic change. The fundamental forces – gravitational, nuclear, and electromagnetic – all 'froze out' as the temperature dropped, fixing the values of $\mathcal{N}$ and ε in a manner that can be considered 'accidental', just like the pattern of ice crystals when water freezes. The number Q, imprinted by quantum fluctuations when a universe was of microscopic size, may also depend on how these transitions occur.

Some universes may manifest different numbers of dimensions, depending on how many of the initial nine spatial dimensions compactify rather than stretch. Even in three-dimensional spaces, there may be different microphysics, and perhaps different values of λ, depending on the type of six-dimensional space into which the other dimensions curl up. Universes could have different values of Ω (which fixes the density and how long their 'cycle' lasts if they recollapse), and Q (which measures how smooth a universe is, and so determines what structures emerge in it). In some, gravity could be so overwhelmed by the repulsive effect of the 'vacuum energy' (λ) that no galaxies or stars can form. Or the nuclear forces may be outside the range of {ε close to 0.007} that allows elements like carbon and oxygen to be stable, and to be synthesized in stars: there would then be no periodic table and no chemistry. Some universes could have been short-lived, and so dense throughout their lives that everything stayed close to equilibrium, with the same temperature everywhere.

And some universes might just be too small and simple to permit any internal complexity at all. I have highlighted one basic number, $\mathcal{N}$, that is exceedingly large – one followed by 36 zeros. Its size reflects the weakness of gravity: very large numbers of particles have to gather together before gravity becomes important – as it does, for instance, in stars (gravitationally bound fusion reactors). It's a straightforward consequence of their size that stars have lifetimes that are enormously long, allowing time for photosynthetic and evolutionary processes to unfold on suitable planets in orbit around them. In Chapter 3 we imagined a universe where $\mathcal{N}$ wasn't as huge as 10^{36} but where everything else (including our other five numbers) was unchanged. Stars and planets could still exist, but they would be smaller and would evolve quicker. They would not offer the stretches of time that evolution demands. And gravity would crush anything large enough to evolve into a complex organism.

The recipe for any 'interesting' universe must include at least one very large number: clearly, not much could happen in a universe that was so constricted that it contained few particles. Every complicated object must contain a large number of atoms; to evolve in an elaborate way, it must also persist for a long time – many, many times longer than a single atomic event.

But an abundance of particles, and a long stretch of time, are not in themselves sufficient. Even a universe as large, long-lived and stable as ours could contain just inert particles of dark matter, either because the physics precludes ordinary atoms from ever existing or because they all annihilate with exactly equal numbers of antiatoms.

THE MYSTERY OF λ

These speculative ideas offer a new perspective on λ, the key number that measures the energy content of empty space. The energy that drove inflation is presumed to have been latent in the vacuum. This means that λ in the remote past was larger by 120 powers of ten than it could possibly be today. In this perspective, it seems surprising that λ should decay away to be so close to zero. There are three very different resolutions of this puzzle.

One is that the microstructure of space (maybe involving a foam-like assemblage of tiny interlinked black holes) somehow adjusts itself to make this so. A second idea is that the decay is gradual, and somehow 'tracks' the density of ordinary matter; it might then not be coincidental that the vacuum should now contribute about the same as the ordinary matter, so that Ω is around 0.3 but the vacuum still stores enough energy to provide the remaining 0.7 that's needed to bring the overall density up to the critical value required for a flat universe.

A third possibility is that there's no fundamental explanation for the smallness of λ in our universe, but that its 'tuning' (like that of our other numbers) is a prerequisite for our existence. We can think of λ as neutralizing the gravity at a particular density; this is what would happen in the static universe that Einstein had in mind when he invented the idea. So, as the universe expands, and the ordinary material gets more diffuse, the density at some stage drops below a threshold and the repulsion starts to 'win' over gravity. Our own universe may have passed that threshold, so that galaxies are already speeding up in their recession from us. But imagine a universe that was 'set up' exactly like ours except that λ was much larger. Then the repulsion would take over much earlier. If this transition had happened before galaxies had formed, then they never would – such a universe would be sterile.

In the multiverse, λ could range over many possible values: these could either be a set of discrete numbers (determined by the way the extra dimensions curled up), or else a continuum of possibilities. In most universes, λ would be vastly higher than in ours. But our universe could be typical of the subset in which galaxies could have formed.

A KEPLERIAN ARGUMENT

The issue of the multiverse might seem arcane, even by cosmological standards, but it affects how we weigh the observational evidence in the current debate about Ω and λ.

Some theorists have a strong prior preference for the simplest universe, with (contrary to the best present evidence) enough intergalactic dark matter to make Ω *exactly* unity, thus implying a degree of tuning in the early universe that was not merely remarkable but absolutely perfect. They're uneasy with Ω being, say, 0.3 and even more by extra complications like a non-zero λ. As we've seen, it now looks as though a craving for such simplicity will be disappointed.

Perhaps we can draw a parallel with debates that occurred 400 years ago. Kepler discovered that planets move in ellipses not circles. Galileo was upset by this. He wrote 'For the maintenance of perfect order among the parts of the Universe, it is necessary to say that movable bodies are movable only circularly'.[2]

To Galileo, circles seemed more beautiful; and they were simpler – they are specified just by one number, the radius, whereas an ellipse needs an extra number to define its shape (the 'eccentricity'). Newton later showed, however, that all elliptical orbits could be understood by a single unified theory of gravity. Had Galileo still been alive when *Principia* was published, Newton's insight would surely have joyfully reconciled him to ellipses.

The parallel is obvious. A universe with low Ω, non-zero λ and so forth may seem ugly and complicated. But maybe this is our limited vision. Our Earth traces out one ellipse among an infinity of possibilities, its orbit being constrained only by the requirement that it allows an environment conducive for evolution (not getting too close to the Sun, nor too far away). Likewise, our universe may be just one of an ensemble of all possible universes, constrained only by the requirement that it allows our emergence. So I'm inclined to go easy with Ockham's razor[3]: a bias in favour of 'simple' cosmologies may be as short-sighted as was Galileo's infatuation with circles.

If there were indeed an ensemble of universes, described by different 'cosmic numbers', then we would find ourselves in one of the small and atypical subsets where the six numbers permitted complex evolution. The seemingly 'designed' features of our universe shouldn't surprise us, any more than we are surprised at our particular location within our universe. We find ourselves on a planet with an atmosphere, orbiting at a particular distance from its parent star, even though this is really a very 'special' and atypical place. A randomly chosen location in space would be far from any star – indeed, it would most likely be somewhere in an intergalactic void millions of light-years from the nearest galaxy.

At the time of writing, the view that our six numbers are accidents of cosmic history is no more than a 'hunch'. But it could be firmed up by advances in our understanding of the underlying physics. More importantly for its standing as a genuinely scientific hypothesis, it is vulnerable to disproof: we would need to seek a different interpretation if the numbers turned out to be *even more special* than our presence requires. Suppose, for instance, that (contrary to current indications) λ contributed less than 0.001 of the critical density, and was thus thousands of times smaller than it needed to be merely to ensure that cosmic repulsion didn't inhibit galaxy formation. This would raise suspicions that it was indeed zero for some fundamental reason. Likewise, if the Earth's orbit had been an *exact* circle (even though we could exist equally comfortably in a modestly eccentric orbit), it could have favoured the kind of explanation that Kepler and Galileo would have preferred, whereby the orbits of the planets were fixed in exact mathematical ratios.

If the underlying laws determine all the key numbers uniquely, so that no other universe is mathematically consistent with those laws, then we would have to accept that the 'tuning' was a brute fact, or providence. On the other hand, the ultimate theory might permit a multiverse whose evolution is punctuated by repeated Big Bangs; the underlying physical laws, applying throughout the multiverse, may then permit diversity in the individual universes.

PROGRESS AND PROSPECTS: A RESUMÉ

Elucidating the ultra-early universe and clarifying the concept of the multiverse are challenges for the next century. These challenges look less daunting if we look back at what has been achieved during the twentieth century. A hundred years ago, it was a mystery why the stars were shining; we had no concept of anything beyond our Milky Way, which was assumed to be a static system. In contrast, our panorama now stretches out for ten billion light-years, and its history can be traced back to within a fraction of a second of the 'beginning'.

Physical probes are, of course, still confined to our own Solar System, but improvements in telescopes and sensors allow us to study galaxies so far away that their light has been journeying towards us for ninety per cent of the time since the Big Bang. We have mapped, at least in outline, most of the volume that is in principle accessible to us, though we suspect that, beyond our horizon, our universe encompasses a vastly larger volume from which light has not yet had time to reach us (and perhaps never will).

We are learning how cosmic structure emerged, and how galaxies evolved, from detailed observations – not only of nearby galaxies but also of populations of distant galaxies that are being seen as they were up to ten billion years ago.

This progress is possible only because of the contingency – in principle, remarkable – that the basic physical laws are comprehensible and apply not just on Earth but also in the remotest galaxies, and not just now but even in the first few seconds of our universe's expansion. Only in the first millisecond of cosmic expansion, and deep inside black holes, do we confront conditions where the basic physics remains unknown.

Cosmologists are no longer starved of data. Current progress is owed far more to observers and experimentalists than to armchair theorists. But in future there will be armchair 'observers'. The results of galaxy surveys, detailed 'maps' of the sky, etc, will be available electronically to anyone who can access or download them. A far larger community will be able to participate in exploring our cosmic habitat, checking their own 'hunches', seeking new patterns, and so forth.

Observations are steadily improving, but our understanding is advancing in a zigzag fashion. There is a sawtooth advance as theories come and go, but the general gradient is upwards. Progress requires more powerful telescopes, and enhanced computer power that permits more realistic simulations.

There are three great frontiers in science: the very big, the very small and the very complex. Cosmology involves them all. Within a few years, the cosmic numbers, λ, Ω and Q should be as well measured as the size and shape of the Earth have been since the eighteenth century. We may by then have solved the problem of the 'dark matter'.

But it remains a fundamental challenge to understand the very beginning – this must await a 'final' theory, perhaps some variant of superstrings. Such a theory would signal the end of an intellectual quest that started with Newton, and continued through Maxwell, Einstein and their successors. It would deepen our understanding of space, time, and the basic forces, as well as elucidating the ultra-early universe and the centres of black holes.

This goal may be unattainable. There could be no 'final' theory; or, if there is, it could be beyond our mental powers to grasp it. But even if this goal is reached, that would not be the end of challenging science. As well as being a 'fundamental' science, cosmology is also the grandest of the environmental sciences. It aims to understand how a simple 'fireball' evolved into the complex cosmic habitat we find around us – how, here on Earth, and perhaps in many biospheres elsewhere, creatures evolved that are able to reflect on how they emerged.

Richard Feynman used a nice analogy to make this point. Imagine you'd never seen chess being played before, then by watching a few games, you could infer the rules. Physicists, likewise, learn the laws and transformations that govern the basic elements of nature. In chess, learning the moves is just a trivial preliminary on the absorbing progress from novice to grand master; by analogy, even if we knew the basic laws, exploring how their consequences have unfolded over cosmic history is an unending quest. Ignorance of quantum gravity, subnuclear physics, and the like impedes our understanding of the 'beginning'. But the difficulties of interpreting the everyday world and the phenomena that astronomers observe stem from their *complexity*. Everything may be the outcome of processes at the subatomic level, but even if we know the relevant equations governing the microworld, we can't, in practice, solve them for anything more complex than a single molecule. Moreover, even if we could, the resultant 'reductionist' explanation would not be enlightening. To bring meaning to complex phenomena, we introduce new 'emergent' concepts. (For example, the turbulence and wetness of liquids, and the textures of solids, arise from the collective behaviour of atoms, and can be 'reduced' to atomic physics, but these are important concepts in their own right; so, even more, are 'symbiosis', 'natural selection', and other biological processes).

The chess analogy reminds us of something else. There is no chance that our finite observable universe, even though it extends ten billion light years around us, can 'play out' all its potentialities. This is because any estimate of how many different chains of events could happen quickly runs into even vaster numbers than we've encountered so far. The number of different chess games, even after only three moves by each player, is about 9 million. There are far more 40-move games than the 10^{78} atoms within our horizon: even if all the material in the universe were constituted into chess boards, most possible games would never be played. And the range of options in a board game is obviously minuscule compared to the variety allowed in nature.

Even simple inanimate systems are generally too 'chaotic' to be predictable: Newton was actually lucky to find, in planetary orbits, one of the few aspects of nature that *are* highly predictable! Any biological process involves tremendously more variety – more branch points at every stage as the complexity unfolds – than a game of chess. If there were millions of Earth-like planets in each galaxy that all harboured life, each one would be distinctive. (Far beyond our horizon, however, there could be a literally infinite expanse, where every possible combination of circumstances could occur – and could indeed be replicated infinitely often[4].) This perspective should caution us against scientific triumphalism – against exaggerating how much we'll ever really understand of nature's intricacies.

A theme of this book has been the intimate links between the microworld and the cosmos, symbolized by the *ouraborus* (Figure 1.1). Our everyday world, plainly moulded by subatomic forces, also owes its existence to our universe's well tuned expansion rate, the processes of galaxy formation, the forging of carbon and oxygen in ancient stars, and so forth. A few basic physical laws set the 'rules'; our emergence from a simple Big Bang was sensitive to six 'cosmic numbers'. Had these numbers not been 'well tuned', the gradual unfolding of layer upon layer of complexity would have been quenched. Are there an infinity of other universes that are 'badly tuned', and therefore sterile? Is our entire universe an 'oasis' in a multiverse? Or should we seek other reasons for the providential values of our six numbers?

NOTES

1. THE COSMOS AND THE MICROWORLD

1 Images depicting the full range of scales in our universe, from largest to smallest, were originally presented by Dutchman Kees Bieke in *Cosmic View: the Universe in Forty Jumps* (John Day, 1957) but were developed further, and achieved widest currency, in a film and book entitled *Powers of Ten* by the office of Charles and Ray Eames, together with Philip and Phylis Morrison. (W. H. Freeman, 1985)

2. OUR COSMIC HABITAT I: PLANETS, STARS AND LIFE

1 An alternative technique being developed is to repeatedly measure the star's position accurately enough to trace its orbital 'wobble'. (Whereas the Doppler technique measures motions along the line of sight, this method detects transverse motions in the plane of the sky.)

3. THE LARGE NUMBER $\mathcal{N}$: GRAVITY IN THE COSMOS

1 This is the mechanical work that must be done to remove an atom from a sphere. It can be thought of as the 'inverse square' force, scaling as (mass)/(radius)^2, multiplied by the distance through which the force acts, which is proportional to (radius). It is also known as the 'binding energy'. It scales as (mass)/(radius), and therefore as (mass)$^{2/3}$ because, for constant densities, (radius) scales as (mass)$^{1/3}$.
2 *The End of Time* (Weidenfeld & Nicolson, 1999).
3 This uncertainty about extreme conditions near the singularity doesn't erode our confidence in the existence of black holes or in our understanding of their external properties. No more does the mystery of

quarks reduce our confidence in the standard physics of atoms, which depends on the behaviour of electrons in orbits on much larger scales.

4. STARS, THE PERIODIC TABLE, AND ε

1 Livio *et al.* (*Nature*, *340*, 281 1989) have computed just how sensitive the carbon production is to changes in the nuclear physics.

5. OUR COSMIC HABITAT II: BEYOND OUR GALAXY

1 According to Einstein's theory, the gravitational attraction depends not on density alone but on [(density)+3(pressure)/c^2]. Leaving out the second term makes a factor of two difference when the pressure of radiation is important. However, we shall see in Chapter 7 that even in empty space there may be some energy. If so, it will have a pressure that is negative (i.e. like a 'tension'). The second term then cancels out the first, and causes a major qualitative change: the expansion actually accelerates rather than slows down. This counterintuitive result is important in the early inflationary universe, and also at present if the energy of empty space (lambda – see Chapter 7) is dominant.
2 There is no mixing between the Sun's centre and its outer layers, so there would be still more helium in the core, because of the spent fuel from the fusion that has kept it shining over its 4.5 billion year history.

6. THE FINE-TUNED EXPANSION: DARK MATTER AND Ω

1 The precise value of the critical density, and indeed some of the other densities quoted here, depend on the actual scale of the universe – something that is only known with 10–20 per cent precision because of the problems of determining the so-called 'Hubble constant'. These issues merit a whole book to themselves. However, I should mention, for the benefit of specialists, that the numbers quoted here correspond to a Hubble constant (in the usual units) of sixty-five kilometres per second per megaparsec.
2 A much more interesting question is whether the inverse-square law breaks down on very *small* scales, or – which is more or less the same thing – whether some extra 'fifth force' comes into play on scales below a few metres. Speculations connected with superstring theories (see Chapter 10) suggest that the extra spatial dimensions may conceivably manifest themselves in this way. Here again, experimental evidence is

meagre, and less exact than we would wish, because gravity is so feeble between laboratory-sized objects.

3 *Less* deuterium when the density is *higher* at first sight seems a perverse result, but it's actually quite natural. The higher the density, the more often the nuclei would hit each other, and the more quickly nuclear reactions would convert hydrogen (one proton) into helium (two protons and two neutrons). Deuterium (one proton and one neutron) is an intermediate product. Not much would survive if the density were high, because the reactions would have gone so quickly that nearly all the deuterium would have been processed into helium; on the other hand, if the density were lower, we would expect more 'fossil' deuterium left over from the first three minutes of our universe's existence. The dependence is quite sensitive, so any reasonably accurate measurement of the deuterium fraction tells us the average density of atoms in the universe.

4 The evidence actually tells us the differences between the squares of the masses of two different species of neutrinos. An earlier version of Kamiokande recorded eleven events due to high-energy neutrinos from the nearby 1987 supernova, mentioned in Chapter 4; an American experiment (in a salt mine in Ohio) recorded eight more. These numbers pleased astrophysicists because they fitted well with what supernova theories predicted.

7. THE NUMBER λ: IS COSMIC EXPANSION SLOWING OR SPEEDING?

1 The successive 'wavecrests' in the light from any atom or molecule are due to its vibrations, which are essentially a microscopic clock. The wavecrests arrive slower when the source is receding and the wavelengths are stretched.

2 From *Nature's Imagination*, edited by J. Cornwell (Oxford University Press, 1998).

8. PRIMORDIAL 'RIPPLES': THE NUMBER *Q*

1 At first sight, this may seem contrary to the statement that *Q* is the same on all scales. However, *Q* is actually measured by the overdensity multiplied by the square of the lengthscale. According to Newton's laws of gravity, the gravitational binding energy at the surface of a sphere depends on mass/radius. However, for spheres of different mass but the same density, mass depends on $(\text{radius})^3$, and so the binding energy varies as $(\text{radius})^2$. So the density fluctuations have smaller amplitude on larger scales.

2 One might wonder why the substructure within galaxies has been erased, whereas individual galaxies survive within a cluster of galaxies (which doesn't become a single 'supergalaxy'). This is because, in the later stages of the hierarchical clustering, the gas is too hot and diffuse to be able to condense into stars. The star formation process is 'quenched' on scales bigger than galaxies.

9. OUR COSMIC HABITAT III: WHAT LIES BEYOND OUR HORIZON?

1 In particular, the intensity measured by COBE at millimetre wavelengths might have been *weaker* than the predicted extrapolation from what had already been reliably determined at centimetre wavelengths. Many processes could have added *extra* millimetre-wave radiation – for instance, emission from dust, or from stars at very high redshifts – and so we would not have been fazed if it had been *more* intense than a black body at these wavelengths. But it would be hard to interpret a millimetre-wave temperature that was *lower* than that at centimetre wavelengths.

2 *The Inflationary Universe* is published by Jonathan Cape, 1997.

10. THREE DIMENSIONS (AND MORE)

1 *Time's Arrow* is published by Jonathan Cape, 1991.

11. COINCIDENCE, PROVIDENCE – OR MULTIVERSE?

1 From *Quarks, Chaos and Christianity*, by John Polkinghorne (SPCK Triangle Press, 1994).

2 From *Dialogues Concerning the Two Chief Systems of the World*, translated by S. Drake (Berkeley, 1953).

3 William of Ockham advanced the view which (translated from Latin) means 'Don't multiply entities more than is absolutely necessary'.

4 As the cosmologist John Barrow has quipped: if this statement is true, it certainly isn't original.

INDEX

4/00

	DATE DUE		